MODERN WORLD NATIONS

Turkey

Zoran Pavlović

Series Consulting Editor
Charles F. Gritzner
South Dakota State University

Philadelphia

Frontispiece: Flag of Turkey

Cover: Tenth-century Armenian church at Lake Van, Turkey.

CHELSEA HOUSE PUBLISHERS

VP, NEW PRODUCT DEVELOPMENT Sally Cheney
DIRECTOR OF PRODUCTION Kim Shinners
CREATIVE MANAGER Takeshi Takahashi
MANUFACTURING MANAGER Diann Grasse

Staff for TURKEY

EXECUTIVE EDITOR Lee Marcott
PRODUCTION EDITOR Megan Emery
ASSOCIATE PHOTO EDITOR Noelle Nardone
SERIES DESIGNER Takeshi Takahashi
COVER DESIGNER Keith Trego
LAYOUT 21st Century Publishing and Communications, Inc.

A Haights Cross Communications ⌐ Company

http://www.chelseahouse.com

First Printing

1 3 5 7 9 8 6 4 2

Library of Congress Cataloging-in-Publication Data

Pavlović, Zoran.
   Turkey/by Zoran Pavlović.
      v. cm.—(Modern world nations)
Includes bibliographical references and index.
Contents: Physical environment—Turkey through time—People and culture—
Government—Economy—Cities and regions—Turkey's future.
   ISBN 0-7910-7916-3
   1. Turkey—Juvenile literature. [1. Turkey.] I. Title. II. Series.
DR417.4.P38 2004
956.1—dc22

                                                              2003028101

Table of Contents

Turkey

1

Introduction

As your adventure in Turkey begins, it will be to your advantage to prepare for the exciting journey. Knowing what to look for, how it got there, and why it is important makes travel much more interesting! Geographers always prepare for their travels to study new lands and their peoples—and each of us becomes a geographer the moment we begin to read about places. As geo-adventurers, we look at the land, but that is only the beginning of our task. When viewing a landscape—any and all visible features of place—we search for answers to three questions: *What* is *where*, *why* is it *there*, and *why* should we *care* about it? *What* includes all physical and human features of a place: its landforms and ecosystems, its homes and communities, and what people are doing as they go about the tasks of daily life. *Where* refers to locations, distributions, and patterns of features in their spatial arrangement on Earth's surface. *Why* explains how features—

whether cities, industries, transportation routes, or other features—are located and how they interact with one another. Finally *why care* helps explain their importance to us and to others. In addition, we are keenly aware of the importance of cultural ecology—how people culturally adapt to, use, and change the places in which they live.

Everything on Earth's surface has its own place and significance. The location of settlements, for example, can serve as a good example of what geographer Harm deBlij calls "the power of place." In the United States, many settlements were located to take advantage of some natural condition. San Diego, New York City, and Seattle, for example, were located and settled because of protective physical features (natural harbors) that favored the development of seaports. Environmental conditions, however, are important only when they are useful to a people. To Americans, ports were essential for trade, fishing, and contact with lands and peoples across the seas. Las Vegas, Nevada, on the other hand, is located in a parched desert landscape, yet it is a booming city. Much of the growth can be attributed to the decision to legalize gambling in a location close to Southern California's major cities. In this way, a city in the middle of the desert became what is now America's fastest growing urban center. Human decisions and geographic location, rather than aspects of the physical environment, influenced this miracle.

Once you understand the previous examples, the same principles can be applied to any other country. Our mission throughout this book is simple—to explore Turkey through a geographer's eyes. Our task will be quite easy, because few countries in the world have been more influenced by location and their ability to take advantage of natural environmental conditions. Let's now apply to Turkey what has been discussed. First, the country's two largest cities are Istanbul and Ankara. Istanbul is a huge, sprawling city with perhaps one of the world's richest histories. Ankara, on the other hand, would have been difficult to find on a map just a century ago. Nevertheless, both cities play a vital role in Turkish society, even though their history and physical landscapes are vastly different.

As you will learn in the following chapters, Istanbul's history spans several thousand years. Initially, it grew as an ancient Greek colony on a strategically crucial water corridor linking the Mediterranean and Black seas. Later, it became an important center within the sprawling Roman Empire. During much of the second millennium, the city grew as the center of the powerful Ottoman Empire. As is true in the U.S. seaport cities mentioned previously, the natural environment played a vital role in the location and subsequent growth of Istanbul. The city was settled on the Bosporus, the narrow strait that connects the Sea of Marmara, and thus the Mediterranean, with the Black Sea. This water link also forms part of the dividing line between the continents of Europe and Asia. With its advantageous location between major bodies of water and two continents, it appears that Istanbul was destined to become a great city.

Ankara, on the other hand, was a sleepy little village of little regional or national significance. During the 1920s, after the creation of the modern Turkish state, however, the government decided to relocate the capital, then Istanbul. Ankara was chosen because of its central location within the country. This location, leaders believed, would have several benefits. It would be more easily accessible than Istanbul to people living in all parts of the country. It was hoped that placing the capital in a remote area would encourage develop-ment in a previously economically stagnant part of the country. An interior location also would make the capital less vulnerable to possible foreign intrusions. Today, the city's population exceeds 3 million, and it continues to grow; the region in which it is located prospers. This form of action was not unique to Turkey, of course. Ankara had a role in Turkey similar to that of Las Vegas in the United States: It was a city deliberately created, with a precise purpose in mind. Again, location was the key to settlement and growth.

These red pines are growing in the sheltered Limani Bay on the southern coast of Turkey. There are many beaches on the Mediterranean coast and it is possible to swim there through the end of the fall.

Now that you know at least several ways in which geographers view the world and its varied people, environments, and places, let us introduce Turkey—one of the world's most interesting lands. To most Westerners, Turkey still represents a place forbidden and distant. It is located in a very complex and often troubled part of the world: the Middle East. Politics and cultures here are often in sharp conflict. For many years, circumstances seemed to work against Turkey. During the Cold War, it shared a border with both the Soviet Union and Iran. It also was located in close proximity to Israel and its often turbulent neighbors. Most recently, the country has been troubled by the conflict in bordering Iraq.

Domestic problems also have added to Turkey's woes. The Kurdish rebellion in eastern Turkey, for example, has not helped the country improve its image as a desirable tourist destination in the eyes of the Western world. Almost anyone who has visited Turkey, though, returns with wonderful memories and can tell amazing stories about their travels and experiences in this fascinating land. It is unfortunate that the news media often project a negative image of Turkey, a situation that leaves outsiders with the wrong impression of the country and its people.

Today, Turkey stands on the brink of becoming a "developed world" country. It has many factors working in its favor as it works to achieve this goal. In size, it is slightly larger than the state of Texas, but it is home to about 70 million people, a population nearly four times larger. It is located at one of the gateways to Central and Eastern Europe. This location, however, can be both a blessing and an obstacle. Proximity to Europe has benefited the country in many respects, but having the explosive Middle East as a neighbor has many drawbacks. It is a land that spans two continents. The largest part of the country—the peninsular portion conveniently called Asia Minor—lies in Asia. That small portion of the country lying west of the Bosporus, Sea of Marmara, and Dardanelles is European.

Aspects of Turkey's historical and cultural geography have combined to create a general attitude among most Turks that their future is essentially European. Although 98 percent of all Turks consider themselves followers of the Islamic faith, for

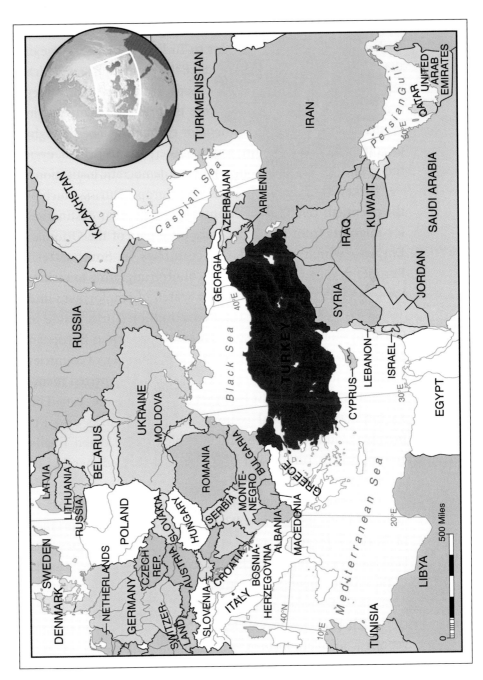

Turkey is located at the gateways to Central and Eastern Europe. It spans two continents; the peninsular portion lies in Asia and the small portion lying to the west is European.

2

Physical Environment

Turkey's location is the primary influence on the conditions of its natural environment. The country is surrounded on three sides by large bodies of water, with the Black Sea to the north, the Aegean Sea forming its western border, and its southern coastal region rising from the Mediterranean Sea. Rugged mountains that cover most of the countryside dominate the physical landscape of the country's interior. They are the result of the same geological processes that caused the formation of surrounding seas millions of years ago. The bodies of water result in a coastline that is about 4,500 miles (7,242 kilometers) long. Most of Turkey's territory lies on the Asian continent; only 3 percent is located in Europe. Despite the country's physical geographic ties to Asia, its primary cultural ties and interests are with the West, particularly Southeastern Europe. Were it not for the small part of Turkey lying west of the Sea of Marmara, the country would not be in a position to think of itself as European.

Mount Ararat is a snow-capped volcanic cone located in extreme northwest Turkey. It is only 10 miles (16 kilometers) west of Iran. Ararat is considered to be a holy mountain due to the references to Noah in the Old Testament. His Ark was said to have come to rest on Ararat following the Great Flood.

LOCATION AND ITS IMPORTANCE

In terms of geopolitics (political geography), Turkey's location has been of considerable importance since its earliest formation as a country. Its position relative to the Bosporus, Sea of Marmara, and Dardanelles gave the country control of this strategically crucial passage between the semi-landlocked Black Sea and the global Mediterranean Sea, and thus the rest of the world. Turkey also shared a political border with the Soviet Union and Bulgaria, both members of the Warsaw Pact. Although the Cold War is over, Turkey's geographic location still plays an important role. For example, a land-surface continental transportation network between Western Europe (the European Union) and the Middle East must cross Asia Minor. Needless to say, the national government is aware of the tremendous economic benefits to be gained once this vital link is established.

Compared with other countries, Turkey is medium in size, covering an area of 301,389 square miles (780,594 square kilometers). Thrace, the small part of the country's territory lying in Europe, occupies an area of 9,065 square miles (23,478 square kilometers). In shape, the country is elongate, stretching slightly more than 1,000 miles (1,609 kilo-meters) in an east-west direction. Its north-south dimension, from the Black to Mediterranean seas, averages about 400 miles (644 kilometers). Only one of its immediate neighbors, Iran, covers a larger territory.

Because of the country's size, one might expect Turkey to have longer boundaries with neighboring countries. It is a peninsula, however, so the country shares administrative borders with only six other countries. Its total length of land boundaries is 1,645 miles (2,648 kilometers). To the west, Turkey shares borders with European neighbors Bulgaria (149 miles, 240 kilometers) and Greece (128 miles, 206 kilometers). Its Asiatic neighbors to the east and southeast, listed in counterclockwise order and including land boundary distances, are Syria (510 miles, 821 kilometers), Iraq (218 miles, 351 kilometers), Iran (310 miles, 499 kilometers), Azerbaijan (5.59 miles, 9 kilometers), Armenia (166 miles,

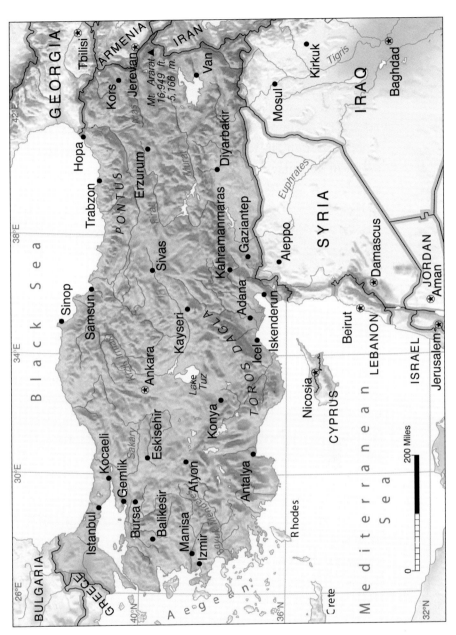

Turkey is composed of two physical areas: the Anatolian Plateau and associated mountains and the coastal regions of hills, mountains, and small plains bordering the Black, Aegean, and Mediterranean seas. Because of the country's size, one might expect it to have longer boundaries with its neighbors, but because it is a peninsula, it borders only six other countries.

267 kilometers), and Georgia (156 miles, 251 kilometers). The length of the border between Turkey and Azerbaijan is not a typographical error. Although it is one of the world's shortest, this boundary is politically and economically important for both countries. A careful look at the map also reveals that almost all islands in the Aegean Sea belong to Greece rather than to Turkey, even though some of them lie just a few miles off Turkey's coast.

THE LAND

To physical geographers, geosynclines are of considerable importance. These huge areas of downwardly warped earth often cover thousands of square miles. They can be responsible for the formation of both land features and depressions occupied by bodies of water. Geosynclines can include deep-sea rifts where tectonic plates—huge "floating" masses of Earth's crust—collide. Such collisions occur over huge areas and often result in the formation of geologically young and growing mountain ranges and bodies of water that occupy one or more adjacent geosynclines.

One of the best examples of these geologic processes is Japan and its immediate area. First, on the eastern side of the Japanese islands, the seafloor almost immediately drops to a depth of more than 15,000 feet (4,572 meters). Second, the process of subduction—one plate plunging beneath another—creates the upward thrust of land and a geologic environment conducive to volcanic and seismic (earthquake) activity. Mountain peaks rise above the sea, forming Japan's nearly 2,000 islands. The adjacent geosyncline has formed a semi-internal body of water, the Sea of Japan, which is separated from the Pacific Ocean. Japan, of course, also has many volcanic peaks and is in constant danger of violent seismic activity.

By now, you might be wondering, "Why are we talking about Japan, when this book is about Turkey?" You must remember that geographers always apply a spatial perspective to their questions, and they learn from examples. Once we understand how a geosyncline system works in one part of the world (by learning from a well-known example), it is much easier to understand other systems elsewhere. Turkey owes much of the look of its land to the same physical processes that Japan does.

Tens of millions of years ago, soon after the dinosaurs vanished at the end of the Mesozoic geologic period, Turkey's land features began to take shape. The African tectonic plate continued pushing northward toward Europe, just as the Pacific plate was doing in the vicinity of Japan. Europe's answer was the formation of a mountain system that begins with the Alps and continues all the way to the Himalayas in southern Asia. Even a brief look at the map shows how important this process was for Turkey.

In Turkey, two east–west trending mountain ranges roughly parallel the northern and southern areas of the Anatolian Peninsula. They are the Pontic (Pontus) Mountains in the north and Taurus (Toros Daglari) range in the south. The Anatolian Plateau lies between the two ranges. As the land elevated to the south and sank to the north, the Black Sea formed as a giant inland lake. Today, it is a saline (saltwater) body of water joined to the rest of the world by a narrow link through the Bosporus. Turkey also has many volcanoes, including the famous Mount Ararat, believed by some to be the site on which the biblical Noah's ark is located. Frequent seismic activity in the region suggests a continuous evolution of the ongoing geologic processes in Asia Minor. In recent years, Turkey has experienced some devastating earthquakes that took thousands of lives.

Mountain Ranges

In the north, the Pontus, or North Anatolian system, rises out of the Black Sea. The name "Pontus" comes from the ancient Greeks, who established many colonies around the Black Sea ("Pont" meaning "sea" in Greek). Today, Turks prefer the name "North Anatolian system." The mountains gradually increase in elevation from several thousand feet in the west to more than 10,000 feet (3,048 meters) in the east. The highest point, reaching 12,900 feet (3,932 meters), is the Kackar Dagi, located not far from the country's eastern border with Georgia. Fortunately, the mountains have not served as a major barrier to traffic between the country's interior and coast. Many rivers flow from the interior highlands to the coast. Their valleys have long served as relatively easy transportation corridors. Northern mountain slopes are forested and are visually more attractive. They receive greater amounts of precipitation because of prevailing winds and their proximity to the Black Sea. Isohyets (lines on a map that connect the places of equal precipitation) sharply follow mountain ridges.

On the southern side of the Anatolian Peninsula, another mountain range—the Taurus (Toros Daglari)—rises out of the sea, creating spectacular scenery in many locations. The range begins near the coastal city of Antalya and extends in an easterly-northeasterly direction, following the Mediterranean Sea to the vicinity of Adana, where the highest peaks reach nearly 10,000 feet (3,048 meters). East of Adana, the Taurus is known as the Anti-Taurus, and the range continues eastward into Iran, where it forms the Zagros Mountains. The ranges merge in eastern Turkey to form the Eastern Anatolian Mountains. There are two major differences between the northern and southern ranges. The first is geologic, the Pontus being predominantly volcanic in origin, whereas the Taurus is formed from folded layers of sedimentary rock. The second difference, closely related to the first, is that because of its porous limestone structure, fewer streams flow through the Taurus.

The area of Cappadocia is southeast of the city of Ankara in Central Anatolia. Volcanic rock formations create an unusual landscape.

Interior Highlands and Plateaus

The region of Anatolia spreads across Turkey's interior to the border with Iran and is the old homeland of the Osman (Ottoman) Turks. These tribes established a monumental empire that lasted for five centuries. Although Turkey's coastline is rather long, its coastal plains are of limited extent. A narrow transitional zone occurs between the coast and the interior highlands and plateaus. Elevations here are lower and precipitation is higher than in the deeper interior. Most of the country, however, is dominated by the relatively rugged terrain of Inner Anatolia—the Anatolian Plateau.

In Anatolia, population patterns reflect the characteristics of the land. Compared with western areas of the country, population densities here generally are low and rural settlement and activities dominate the landscape. The Pontus range often blocks air masses from the north, making the plateau much drier than the coasts. Aridity limits agricultural production. This part of Turkey is best used as grazing land, and it has long been home to pastoral nomads. Although the country has no true deserts, several areas of sand dunes do exist in Anatolia. Scientists believe they were formed by the process of desertification—the creation of desertlike conditions by human activity, including overgrazing by livestock. Most of the terrain is quite rugged. Numerous volcanic cones are present, particularly in the central and eastern parts of Anatolia. Here, the highest peaks have permanent snowcaps. They include Mount Ararat, 16,946 feet (5,165 meters); Suphan, 14,540 feet (4,437 meters); Resko, 13,671 feet (4,167 meters); and Cilo, 13,510 feet (4,118 meters). In the interior, a number of large lakes and reservoirs occupy the many basins formed on the plateau surface.

In eastern areas, toward the Iranian border, the results of volcanic activity are visibly represented in the landscape. For millions of years, volcanoes spewed huge amounts of lava and volcanic ash, creating deep layers of igneous rocks. Such layers often reach depths of several thousand feet. Turkey's largest lake, Lake Van, can attribute its existence to volcanoes. Drainage was blocked by a huge lava flow, and water filled a basin behind this natural dam.

Toward the south and Turkey's border with Syria, Anatolia's terrain becomes hilly, with elevations dropping to an average of 1,500 feet (457 meters). This zone is known as the Arabian

Akdamar Island stands in Lake Van, which is in the eastern part of the country. Lake Van is the largest lake in Turkey and was created when water filled a large basin behind a natural dam that was caused by volcanic activity.

Platform, because southward its elevation continuously decreases in the direction of the Arabian Peninsula and the Arab-speaking world.

Coasts

Although Turkey's coastal plains represent a rather small part of its territory, they are of vital importance to the country and its people, particularly in terms of their economic role.

Except in the northwestern part of European Turkey, coastal plains are almost nonexistent in the Black Sea region. Steep mountain slopes dominate the landscape, leaving only a thin corridor between them and the sea. The only exceptions are those areas where rivers flowing from Anatolia enter the sea. The largest such river is the Kizilirmak (Red River); the city of Bafra is located on its significant delta. Fertile soils of this rich alluvial (stream-deposited) plain support a productive agriculture. A relatively mild and pleasant climate and attractive coastal scenery combine to make this one of Asiatic Turkey's most densely populated areas.

In the Mediterranean coastal zone, conditions are even more drastic. Except in the vicinity of Adana, close to the Syrian border, where the Seyhan River creates a broad and fertile alluvial plain, the Taurus rise directly from the sea. The settlements there were established on mountain slopes, because flat coastal plains simply do not exist. Most of southern coastal Turkey's population is found in and around Adana, one of the country's most populated cities. This urban center sprawls across the lowlands created by deposition from the Seyhan River, which by a wide margin is the country's largest alluvial plain.

Coasts bordering the Aegean Sea and Sea of Marmara were formed by land thrust upward along various fault lines. These faults also formed the water passage between the Black and Agean seas. Distance from the sea to the closest mountains in the Aegean coastal region is greater than in the rest of Asia Minor, leaving more open space for settlements and economic activities. It is not surprising that population distribution here reflects the importance of the physical environment. Several rivers flowing from Anatolia into the Aegean Sea have created broad alluvial plains with fertile farmland. The two most important are the Gediz, which enters the sea north of Turkey's third largest town, Izmir, and the Buyuk Menderes, which lies to the south of the city. People living in this area have always appreciated its adequate precipitation and good living conditions, and it was the location of the many important ancient Greek colonies.

CLIMATE AND ECOSYSTEMS

Weather and climate are perhaps the single most important nonhuman factors in the creation of distinctive landscapes. Turkey's landscape is no exception. Climate is a multiple-year average of weather conditions (weather is the current atmospheric condition). It is a primary influence on ecosystems—the composite physical landscape of a region, including its vegetation, soils, animal life, and water features. In addition, ecosystems are very important to humans because they offer various potential methods for land use.

Turkey's Aegean and Mediterranean coastal regions enjoy a Mediterranean climate. This climate, judged by many to be the world's most pleasant, offers long, warm summers and short, mild winters. Extremes are rare, and snow seldom falls except at high elevations. The climate is unique in that winter is the season of precipitation, and summers are often marked by severe drought. Vegetation is similar to that of southern coastal California: short grasses, small bushes, and scattered trees. Because of centuries of cutting, stands of woodland are few and small. An exception is on the northern slopes of the Pontus Mountains, the wettest area of Turkey. Here, dense stands of woodland make this region the center of the nation's lumber industry.

Population distribution and density in Anatolia generally reflect climatic conditions. In the interior, a dominantly rural population, whose primary economic activity has always been agriculture, is highly dependent on climatic conditions. Western Anatolia receives more precipitation from the limited, yet crucial, penetration of Mediterranean air masses. Therefore, it is more densely populated than the eastern part of the peninsula. In general, Anatolia experiences hot and dry

The historic town of Kalkan is on a harbor on the southern, or Mediterranean, coast of Turkey. This ancient harbor is at the feet of the towering
Taurus Mountains.

summers and cool to cold winters, and it always lacks enough
moisture to completely satisfy residents' needs. It is not
unusual for temperatures to exceed 100 degrees Fahrenheit
(38 degrees Celsius). Winters, on the other hand, are mostly
cold and dry and can bring severe conditions when temperatures drop to bone-chilling lows. The lowest temperature ever
recorded in Anatolia is just under−30 degrees Fahrenheit
(−34.4 degrees Celsius).

In Eastern Anatolia, where elevations are higher, residents experience more precipitation (the majority of it in the form of snow), which increases with elevation. Here, it is not unusual for snow cover to last as long as three months. Higher volcanic peaks are permanently mantled with snow at elevations higher than 12,000 feet (3,658 meters). Generally speaking, most of Turkey's natural environment suffers from inadequate precipitation. Vegetation, therefore, is well adapted to conditions of scarce moisture. Short grassland, brush, bushes, and wild flowers dominate the flora. With few exceptions, large expanses of woodland are lacking.

Turkey was an important center of both early plant and animal domestication. With more than 80,000 different animal species, the country has an abundance of fauna. Many different mammal species, including sheep and goats, were originally domesticated in Asia Minor.

WATER RESOURCES

Turkey is more fortunate than other countries in the Middle East. Even so, water represents one of the main environmental issues in this country. Southwestern Asia's nations often experience difficulties with water management in years with low precipitation. Turkey, however, has an advantage over its southern neighbors. Of primary importance in this issue are the eastern Anatolian highlands, an area where three major rivers begin their flow toward the sea. The Tigris and Euphrates (Dicle and Firat in the Turkish language) are well known worldwide, because they were the source of life during the dawn of civilizations in ancient Mesopotamia (today's Iraq). The importance of the third river is appreciated only by those who know more about Turkey, its past, and its present. Kizilirmak is Turkey's longest river, and from the beginning of its flow to the delta, where it meets the Black Sea, this artery of life stays entirely inside Turkey's national boundaries.

Turkey's rivers drain into the several large bodies of water and some smaller inland reservoirs. The Tigris and Euphrates flow southward and ultimately join to form the Shat-Al Arab, flowing into the Persian Gulf. Most smaller streams in southeastern Turkey are tributaries of these two rivers. Before crossing the Syrian border, the Euphrates meanders through the Anatolian highlands for 688 miles (1,107 kilometers), making it Turkey's second longest river. The Euphrates owes its source to snow accumulations at elevations higher than 9,500 feet (2,896 meters). The Tigris, which also has its source in the Anatolian (Taurus) highlands, accepts the rest of the tributaries in the region, but its length through Turkey—only 281 miles (452 kilometers)—is much shorter than that of the Euphrates. This is enough, however, to make it the sixth longest river in Asia Minor. The Tigris, too, receives most of its water from melting spring snow. Neither river can depend on different forms of precipitation because of the dry climate throughout the rest of year. Water from both rivers is used for agricultural irrigation, and the flow of each is harnessed for the generation of hydroelectric power.

The 715-mile-long (1,151-kilometer-long) drainage system of the Kizilirmak covers most of central Anatolia. One of the earliest known civilizations in Asia Minor, the Hittites, whose zenith was between the eighteenth and thirteenth centuries B.C., organized their life around the Kizilirmak. Today, water from the river is used extensively for irrigation in what is the driest part of the country. Both the Kizilirmak and Turkey's third longest river, the Sakarya (491 miles, 790 kilometers), drain into the Black Sea. Because both rivers severely erode the lands through which they flow, extensive deposition of alluvial materials occurs at their mouths. Several rapidly expanding urban areas depend on the rivers for their development, but both the amount and quality of the water available in this part of the country are limited. For a nation of 70 million people that needs to better distribute its population, further settlement expansion into the interior will depend on better quality control of those streams flowing from the Anatolia into the Black Sea.

The Mediterranean and Aegean seas also receive water from several major rivers. Here, as in the rest of the country, the distribution of precious river water plays an important role. Precipitation contributes more moisture here than elsewhere in the country, and it is more equally distributed throughout the year, but this does not satisfy the demands of the region's growing economy. In Anatolia, a significant amount of water is lost to evaporation and transpiration (passage of water vapor from a living body, in this case, from plants) during summer months. As a result of irrigation, farmers must irrigate. This use of water, however, limits the amount available to coastal towns bordering the Aegean Sea.

The Buyuk Menderes River is the most important river flowing into the Aegean Sea from Anatolia. Similar to that of the Kizilirmak, the role of the Buyuk Menderes was vital to early civilizations, in this case ancient Greece. In fact, the geographic term "meander," describing a stream that snakes through the countryside instead of flowing straight, is derived from the Greek name for this river. The river is only 135 miles (217 kilometers) long, but it drains an area of 10,000 square miles (25,900 square kilometers). Other significant streams are the Gediz, Seyhan, and Ceyhan. The latter two create the previously mentioned fertile alluvial plain in the southern coast in the vicinity of Adana.

Even the Caspian Sea receives drainage from Turkey. East Anatolia's mountains are the origin for that water, most of which flows through the Aras's (Araks) system. The Aras River's upper flow is entirely in Turkey. It later forms a border between Turkey and Armenia before finally continuing through Iran and Azerbaijan to flow into the Caspian Sea.

A number of Turkey's small rivers flow into inland lakes or reservoirs. Most of the country's lakes are located in the east, where precipitation is higher throughout the year. The country's largest lake, Lake Van, occupies an area of 1,430 square miles (3,764 square kilometers). Although several streams bring fresh water into Lake Van, there is no outflow; hence, the lake is saline. This lake is beautiful because of its clear, blue water. Historically, it also has played an important cultural role. In Armenian tradition, the vicinity of Lake Van is the homeland where Armenians originated and evolved their ethnic identity. Lakes in western Turkey are mostly situated in basins of interior drainage; therefore, their salinity levels are high. One of the best examples is 625 square mile (1,619 square kilometer) Lake Tuz, located southeast of Ankara in the Central Massif.

NATURAL HAZARDS

Each Turk is aware that a potentially deadly catastrophe can hit at any time. Some countries, such as the United States, experience a variety of environmental hazards. Turkey, on the other hand, is known for only a single catastrophic hazard—earthquakes. Throughout the country's recorded history, these seismic events have struck frequently, often with devastating outcomes. In recent years, thousands of people have died from this unpredictable force of nature. In addition, economic losses from the damage and the funds needed for reconstruction have amounted to billions of dollars. In one of the latest catastrophes, at the Sea of Marmara in 1999, about 17,000 people were killed and the economic loss was estimated to be about $10 billion.

The impact of natural hazards is magnified by cultural factors. For example, many buildings are very poorly designed and are built with substandard materials. Residents are extremely vulnerable to injury or death when buildings collapse during an earthquake. Weak local governments result in poor planning, which, in turn, contributes to extensive damage, particularly in poor urban areas, where most damage is done. Today, people are attracted to city life and are moving into fast-growing, overcrowded settlements that pose great danger. Such problems are not limited to less developed countries, however. Many areas of the United States, for example, support huge populations living in hazard zones.

ENVIRONMENTAL ISSUES

Environmental preservation is a growing trend in Turkey. The country has many national and natural parks, as well as other preserved areas. About 15 million visitors, both domestic and international, enjoy Turkey's natural treasures. Wetlands host a huge variety of species and are of particular interest to international ecological organizations because of their number and size. Carefully regulated legislation designed to preserve existing species and conditions protects a large number of animal and plant species. Turkey is blessed with a geographic location that has long served as a corridor for the migration of birds and animals that have contributed to the country's diversity of fauna.

3

Turkey
Through Time

Reaching into the past to explain the present and future might seem to be a task for historians. Geographers, however, also employ the historical method when explaining past and present conditions of place and culture. In addition, unlike most historians, geographers do not limit themselves to written documents; rather, they use any and all sources of historical information while attempting to analyze aspects of material and nonmaterial culture. In studying the way of life in a particular location, for example, a geographer will often seek historical linkages or connections with other locations. We are all aware that the exchange of ideas flows from place to place as well as from time to time by a process called diffusion. Simply stated, a great idea does not become great if it stays in one place. Imagine, for example, that a member of a traditional culture in equatorial Africa is developing the most sophisticated alphabet ever created, but its use is limited to several villages in a remote tropical forest for generation after generation. How would it affect the rest of the world? It would not. Looked at in another way, how did the individual get an idea to create a new alphabet in the first place? Did he or she receive any teaching from somebody else prior to invention? Where did the individual get the idea of writing and the need for an alphabet? In any circumstance, diffusion must take place to increase the distribution, use, and importance of culture traits. Accordingly, cultures residing on the corridor(s) of diffusion will benefit the most.

Throughout human history, Asia Minor was exactly such a place. Its location placed it on the crossroads of civilization on the old-world landmass of Asia, Africa, and Europe. It was not by accident that many biblical stories took place in this region. Legend even has Noah's ark landing on Mount Ararat after surviving the Great Flood. The land (place) of the contemporary Turkic state can be described as an interstate highway rest stop with a beautiful view of a nearby national park. Many drivers keep driving, but some decide to stop and perhaps even camp for a while. This is exactly what happened in the settling of Asia Minor. Some invaders kept going, whereas others decided to stay longer and establish great civilizations, which would eventually be replaced by others. In short, this is the (hi)story of Asia Minor.

EARLIEST HISTORY

Turkey is of particular interest to scientists attempting to discover earliest organized settlements. Many now believe that the southern shore of the Black Sea might hold the key to explaining when the biblical event described as the Great Flood occurred. If true, this also would help explain the biblical account of Noah's ark being on nearby Mount Ararat. Recent discoveries of settlements on the floor of the Black Sea not far from the coast provide a clue to the flood As the post–Ice Age global sea level rose (during the last ice age, which ended about 12,000 years ago, the sea level was more than 300 feet [91 meters] lower than it is today), water suddenly began to flow into the basin now occupied by the Black

Sea. Human settlements—including fields, pasturelands, and villages—were inundated in what must have been the greatest catastrophe in history up to that time. This version of the Great Flood is of a much smaller magnitude than that described in the Bible. It was, however, a major event, and its story certainly would have been passed down from generation to generation, eventually finding a place among Semitic peoples to the south. From there, the event entered into the biblical Old Testament. What is known, based on archeological evidence, is that at the dawn of civilizations in Mesopotamia and ancient Egypt, Asia Minor was already widely settled. Furthermore, some sites date back more than 6,000 years. This suggests that settled communities evolved here at very early stages of the agricultural revolution.

More evidence, from the third millennium B.C., clearly points to civilization existing along the shores of the Aegean Sea and the Sea of Marmara. Historians date the ancient city of Troy back at least to that period. To geographers, this does not seem surprising, considering the town's location—on the shortest route between Europe and Asia. Perhaps Troy served as one of the "rest stops" on this important route. At about the same time, Anatolia's population appears to have been of a stock other than Indo-European (the linguistic stock that is the root of most contemporary European tongues). Hittites were the first known Indo-European tribe to establish a major presence on the peninsula. They were settlers from the east who replaced the preexisting Hurrians sometime during the second half of the third millennium B.C. During the following centuries, the Hittites slowly developed a civilization that, at its zenith, was one of the ancient world's greatest powers.

RISE AND FALL OF THE HITTITE KINGDOM

Hittite history is in some respects more interesting than and certainly different from that of other Middle Eastern civilizations.

The ancient ruins of Hattusas are near the present-day city of Kayseri. These storage jars were left by inhabitants thousands of years ago. They are evidence of the capital city built by the Hittites during their settlement in the area from the 18th to the 19th centuries B.C.

It was not until archeologists more carefully explored and examined the ancient ruins at Hattusas (near the Turkish city of Kayseri) that they realized they were handling something of great significance. To gather more details, scientists began studying stories of the Old Testament. Previously, many scholars believed that the Hittites were little more than products of overly fertile imaginations of ancient writers. Biblical stories often mentioned the Hittites and their confrontations with other powers, yet such tales generally were disregarded as being nothing more than ancient fiction. Further investigation of Egyptian manuscripts, however, revealed that such connections had, indeed, existed. It became clear that the Hittites had thrived for many centuries.

The Hittites came from the east during the nineteenth and twentieth centuries B.C. and established a kingdom in Central Anatolia. They built their capital at Hattusas and spread from there over the rest of Asia Minor, creating a powerful confederation. The key to their success was in acceptance of key cultural traits from other peoples. They integrated these ways of living into their own lifestyle, which continued to develop to higher levels. Hittites, for example, were among the first to adopt the use of iron. Iron-made weapons and fast chariots played an important role in the victories of Hittite kings.

As a curiosity, some Hittite names are perhaps the most difficult to spell among all ancient cultures. Some of them, such as Suppilliuma or Hattusili, would be great candidates for a spelling bee! One of the reasons it took so long for scholars to learn about the Hittites was because of their inability to translate the Hittite alphabet. Linguists usually can decode an alphabet quickly if they know the language. In this case, however, their hieroglyphic alphabet and spoken tongue were unfamiliar. Finally, researchers discovered bilingual documents, which helped them translate the language and therefore learn more about the Hittites.

In the early thirteenth century B.C., the famous Battle of Qadesh (from the Syrian city, Qadesh) was fought between pharaoh Ramses II and the Hittites for domination over the entire Middle East. The vicious battle ended without a victor. Both armies retreated. This event marked the beginning of Hittite decline after five centuries of dominance in Asia Minor. Within a century, the Hittite confederation declined under weak rulers and was overrun by others.

ASIA MINOR DURING THE FIRST MILLENNIUM B.C.

Not long after the Hittites disappeared from the world scene, another influential kingdom—the Phrygian—was established in Asia Minor. Originally, the Phrygians belonged to the "Sea People," a group of aggressive tribes that had spread all over the eastern Mediterranean. Their presence was recorded in documents from Greece, Egypt, and elsewhere. The Philistines (of the biblical story of David and Goliath) were one such tribe. Whereas Phrygians controlled the west and central regions of modern Turkey between the eleventh and seventh centuries B.C., the southeast was under the growing influence of another group, the Assyrians. From their capital in Nineveh (northern present-day Iraq), Assyrians spread to become one of the greatest ancient empires. Their area of dominance in Asia Minor was limited to the east and southeast, however, leaving Phrygians to control the west. Every kingdom in history experienced a rise and ultimate fall. The fall of the Phrygians was associated with an anecdotal character, King Midas, who became famous for his "magic touch" that turned everything to gold. He was also described as having "ass [donkey] ears." Because of these and other stories about the mythical (rather than the historical) Midas preserved in Greek mythology, the Phrygians live on today in the pages of history. The actual Midas was an influential ruler who just did not have much luck.

After Midas lost his kingdom to foreign invaders, Lydia rose as a prominent force from the ruins of the Phrygian kingdom. Lydians controlled western Anatolia for more than a century, but their downfall happened quickly. The end of Lydia is associated with a wonderful story about human nature. During the mid–sixth century, Persia was developing into a regional power. In an attempt to prevent its expansion, Lydian King Croesus was considering a campaign against the Persians. Before starting the attack, he decided to consult the Delphi oracle for advice. After receiving the oracle's answer, that if Croesus started the campaign, one great empire would be destroyed, he decided to attack Persia. After the war was over, as the Delphi oracle predicted, one great empire was, indeed, destroyed—Lydia.

GREEK COLONIZATION

During the latter part of the sixth century, numerous Greek colonies appeared on the shores of the Aegean and Black seas. Asia Minor was an attractive destination for ancient Greeks because of its proximity to Greece. It also provided fertile alluvial plains. These rich lands could be colonized and developed agriculturally. A number of colonies, such as Pergamum, Efes, and Milet, became leading Greek towns and important centers of trade and learning. The famous classical writer Homer, who wrote the epics *The Iliad* and *The Odyssey*, was born in the Aegean city of Smyrna. Today, Smyrna is known by its Turkish name, Izmir.

After conquering the rest of Asia Minor, the Persians continued their military campaign not just against the Greek colonies but also against Greece itself. After several successful episodes, the huge Persian force lost in a naval battle at the island of Salamina. Thus, in the early fifth century, the Persians finally retreated from European parts of the Greek colonized lands but continued to be a dominant force in southwestern Asia. Only when Alexander the Great expanded Macedonia into Asia in the 330s B.C. were the Greeks finally freed from the Persians. Alexander's rule was magnificent, yet tragically short-lived because of his premature death from disease at age 33. Within only several years, Alexander had created one of the greatest empires in history. It spread from southwestern Europe to India. He appreciated knowledge and learning, and many cities thrived under his rule. After his death, however, the empire, including Asia Minor, fell into turmoil and decline. Lands included in the once powerful and united empire fell to the rule of Alexander's generals, who divided the former

This stone lion guards the funerary temple of Antiochus of Commagene at Nimrud Dag near Eski Kale in Anatolia. Antiochus ruled from 69–34 B.C., in a kingdom north of Antioch, which is modern-day Antakya.

Macedonian Empire among themselves. Anarchy and conflict plagued the region until it became a part of the Roman Empire several centuries later.

EVOLUTION OF THE EASTERN ROMAN EMPIRE

By the end of the first century B.C., present-day Turkey was absorbed into the powerful and expanding Roman Empire. This rule would continue for more than a thousand years until, finally, Ottoman (also called Osman) Turks freed Asia Minor from Roman control. The Roman Empire, however, introduced many positive changes wherever it expanded. The cliché "All roads lead to Rome" is a legacy of their excellent transportation systems. Perhaps their greatest legacy was the introduction of the Roman legal system, approach to administration, economic regulations, and other institutional strategies. One extremely important change was that the Roman Empire was highly cosmopolitan in nature, more so, perhaps, than any other culture of its time. Even the residents of remote rural areas felt as if they were part of one giant entity. Such an attitude helped preserve the Empire's territorial integrity for many centuries.

By the fourth century A.D., political changes directly affecting Asia Minor began to sweep the peninsula. First, the empire was divided into western and eastern segments, with Anatolia becoming the flagship of the Eastern Roman Empire. Second, the capital was moved to the city of Byzantium. Emperor Constantine decided to relocate the capital, moving it from Rome to a Greek-speaking area. With the change, Byzantium became Constantinopolis (Constantinople). This event was perhaps the single most important decision in the history of that region: Constantinople would later develop into the world's greatest city. Even today, under the name Istanbul, this remarkable metropolis is home to as many as 10 million residents. The first settlement known to exist at the site of present-day Istanbul was a small colony established in the sixth century B.C by settlers that left the Greek city of Megara. In order to honor their ruler, settlers named the colony with his name—Byzantium.

For reasons unknown, the name Byzantium (and Byzantine Empire) became synonymous with the Eastern Roman Empire for the period between the fourth century A.D. and the empire's decline in the mid 1400s, but the distinction appears only in the works of modern historians. In this context, it is very important to realize that citizens of the Eastern Roman Empire regarded themselves exclusively as Romans. Even though they spoke Greek, they maintained Roman traditions. They never used the term Byzantium to describe their homeland. As is true of so many historical myths, however, it is almost impossible to correct this mistake in history textbooks.

Fierce Germanic tribes finally dealt a deathblow to the Western Roman Empire in 476, although lands controlled by Constantinople continued to flourish. The rest of Europe sank into the Dark Ages, but Constantinople thrived as the world's greatest city, both in population and in terms of its importance. At its historical peak, almost 500,000 people lived beyond the city walls in the immediate surroundings, whereas the residents of Paris could not account for more than 10,000. The following centuries would see many influential rulers in Constantinople but also a gradual shrinkage in the Empire's size and influence.

THE WORLD'S GREATEST CITY

The first of many powerful leaders of the Eastern Roman Empire was Emperor Justinian I (527–565 A.D.), who consolidated the territories and regained some lost provinces. Although Justinian I was a skilled ruler and military strategist, he is better known for his architectural contributions. During his reign, Constantinople was restored to its full glory. Many public buildings, both religious and secular, were built to profoundly alter the city's landscape.

One of the most impressive of these buildings was the Hippodrome. This huge structure, used for chariot races, could accommodate 40,000 spectators. A network of nearly impenetrable walls was constructed around the city to protect it from invaders. Although builders erected many monumental structures during Justinian's reign, one stood above all—the cathedral of Aya Sophia. Finished in 537 as the landmark of the Empire's power, Aya Sophia was one of the largest and most beautiful religious buildings in the world. Although nearly 1,500 years old, its unsupported dome, reaching almost 200 feet (61 meters) in height, is still the world's largest and one of the highest! The beautiful structure served as an impressive symbol of the Empire's power over the known world. Throughout its millennium of independent existence, the Eastern Roman Empire was a respectable military factor in the region. Even so,

The Aya Sophia (Hagia Sophia) was finished in 537 A.D. under the rule of the Roman emperor Justinian and it is a landmark of the Roman Empire's power. It was once the world's largest Byzantine church. In 1453, Constantinople (present-day Istanbul) fell to the Muslim conqueror Mehmet II and this church was converted into a mosque. In 1935, it was converted into a museum.

its diplomatic skills often produced more successful results. One of the methods used to "convince" rebellious rulers at the periphery that rebellion was useless was to invite them to Constantinople. Once they witnessed the city's enormous wealth and power, the greedy warlords elected to join, rather than fight, the system. This approach also was successful in the Eastern Roman Empire's foreign policy. Many "pagan" leaders were converted to Christianity after they were introduced to Constantinople.

By the seventh century, the Empire began to experience political difficulties with a growing power to the south. Arabs, and their new and militant Islamic faith, were beginning to expand from the Arabian Peninsula. In less than a century, they controlled lands extending from central Asia to the Atlantic shores of northern Africa. Territories were lost and often regained as the Eastern Roman Empire experienced a sequence of stronger and weaker emperors. Throughout this period, however, Asia Minor remained an integral part of the empire.

The Crusades—armies of Western European Christians seeking to the rid the Holy Land of Islamic influence—started at the end of the eleventh century and continued well into the thirteenth century. These invasions brought many changes to the region, including the end of Constantinople's "Golden Age." From the very beginning, invading crusaders and the region's emperors were often engaged in hostilities. Finally, in 1204, Crusaders invaded and occupied Constantinople, establishing the Latin Kingdom, which lasted for the next half-century. In 1261, the city was liberated, but the Empire was just a weak remnant of what it had once been. To make things worse, another military power was growing on the rolling hills of Central Anatolia: the Ottoman (Osman) Turks.

THE TURKS

Turks have had a presence in Asia Minor for 1,000 years. As an ethnic group, scientists believe that the earliest Turkic tribes originated in present-day Russia, Mongolia, and northern China. Their language is not related to the Indo-European group. Rather, it is a Ural-Altaic tongue, which ties the Turks linguistically to much of central Asia, to many ethnic groups in the Caucasus region, and, in Europe, to Finns and Hungarians.

During the sixth and seventh centuries, the Eurasian land-mass was experiencing a wave of eastward to westward migration. Many Germanic, Slavic, and other tribes were moving west-ward into Europe after being pushed from Asia by stronger tribes. The movement is generally believed to have originated in the fourth century, when another Turkish tribe, the Huns, led by their famous leader Attila, stormed Europe and triggered the migrations. At that time, Turks were divided into many different tribes. They were nomads who began slowly moving toward the west, first into central Asia and later westward to the Middle East. In central Asia, they encountered the Islamic religion. Slowly, they began converting from their traditional forms of animism to this new and rapidly spreading faith. As you will learn later in the book, Islam played a very influential role in the region, and perhaps the most important part of this role was the evolution of the Turkish Empire.

SELJUK AND OTTOMAN (OSMAN) TURKS

Being skilled and fierce fighters, Turks were in demand for military service throughout the region. Had that not been the case, the Turks might never have migrated into the Middle East to ultimately become one of the greatest nations of their region and era. One tribe was known as Seljuk Turks (Turkic tribes were named after their leaders). During the ninth century, this tribe became the mercenary force for the caliph of Baghdad. Their military role was to protect the caliph's lands against possible invaders.

Occasionally, servants were known to overthrow their masters, and this is precisely what happened with the Seljuk Turks and their rulers. In 1055, they occupied and took control of Baghdad. This event ended the era of Abbasid caliphs and established the Seljuk Turks as the dominant force in Mesopotamia. This location placed them right at the gate of Anatolia, and they gradually began pushing into Asia Minor. In 1071, after a major victory at Manzikert, in Eastern Anatolia, against forces sent by Constantinople, the Seljuks were in Asia Minor for good.

After the beginning of the first Crusaders' campaign in 1097 and their occupation of Jerusalem in 1099, however, the

Seljuk Turks temporarily lost their lands. The Crusades, which Pope Urban II called for after the Seljuks occupied the Holy Land, eventually reduced their influence and power. Nonetheless, the Seljuks were able to retain control of a majority of their lands, including large portions of Asia Minor, into the thirteenth century. As history always witnesses, though, one power can quickly replace another. In the Middle East of the thirteenth century, that power was the Mongols. In just a few decades, the Mongols—fierce warriors from central Asia— spread terror throughout much of Eurasia. In so doing, they carved out the largest empire the world has ever known. Lands under their control spread from the shores of the Pacific Ocean to Poland and from frigid Siberia southward to the tropical shores of the Indian Ocean. They quickly overran the Seljuk Turks and captured Baghdad, which ultimately left the lands previously controlled by the Seljuks without a governing power. From this temporary anarchy arose what became the best-known Turkic tribe, the Ottoman (Osman) Turks.

At this point, another historical misconception deeply rooted in textbooks must be clarified. This is not only a matter of terminology but also a matter of understanding the basic differences in the Ottoman Empire's genesis and evolution. You have already learned that Turkic tribes were identified by and named for their leaders. One of the minor tribes in Anatolia during the thirteenth century was under the leadership of Osman (1284–1324), a founder of the dynasty later known as the Ottoman dynasty. Logic calls for that dynasty to be called Osman, or Osmanli, after its first important leader, but in Western textbooks, this correct nomenclature never took hold.

Rather than following the usual practice in naming an ethnic group, the Osmans became identified with the term "ottoman," which, according to every major dictionary, is a sofa, or divan! Once you know this detail, it will become a bit clearer where the mistake originated. Although Ottoman is synonymous for the word "divan," it is also a term for a political institution in some Muslim countries, including the sultanate created by Osman's descendants. "Divan" was a form of assembly that included the council of ministers, the empire's highest political body. How, then, did an ethnic group become identified with a political institution and a piece of furniture? Why, in the West, do we call them Ottoman Turks, instead of Divan Turks, if following the logic?

The answers to these questions are not complicated once one realizes that the word "ottoman" is Germanic in its origin. This means it had to be brought to the Western world by someone who had direct contact with Turks, in this case the German-speaking Habsburg (Hapsburg) Monarchy that for centuries formed a buffer zone between the Turks and the rest of Europe. For several hundred years, the two empires fought numerous wars against each other. No one in the West had closer ties to the Turks than did the Habsburgs. This explains why the word "divan" was replaced with "ottoman," simply in order to convey the same meaning using the common German term. Once it became the common term in the West, it became difficult to correct. If this seems complicated, you must remember that it took nearly 500 years for Native Americans to convince people that they are not related to the inhabitants of the Indian Peninsula, or East Indies (hence, Indians). In this book, to avoid confusion, "Ottoman" is used in reference to the empire, whereas "Osman" refers to the ethnic group referred to as the Osman Turks.

FALL OF CONSTANTINPOLE AND CREATION OF AN EMPIRE

Under Osman's leadership, troops began to control other areas in Asia Minor. Soon the Turks would overrun all lands once part of the Eastern Roman Empire. Constantinople finally fell into Turkish hands in 1453, after a long siege and betrayal from inside the city's walls. During the fourteenth century, Osman Turks firmly established their presence. Christian troops were defeated on several occasions, the biggest victory being at Nicopolis (1396) in present-day Bulgaria, where the Crusaders were soundly defeated. This victory opened the doors for westward expansion. For 300 years (until the end of the seventeenth century), Turks posed a threat to the rest of Europe. The Mongol Tamerlane's (Timur) invasion of Anatolia in 1402 and various dynastic battles among throne contenders posed periodic obstacles, but even these setbacks did not slow down the empire's expansion throughout the fifteenth and sixteenth centuries.

The key to Turkic expansion was the successful organization and integration of other ethnic groups. Initially, because they were a minority population, the Osman Turks' empire-building process had to rely on assistance from other ethnic groups. Islam's role was crucial in this process. At that time, ethnic differences were distinguished primarily on the basis of religion, which made Islam attractive to many for, if nothing else, economic purposes. Gradually, through cultural integration, the ethnic group recognized today as "Turks" was formed. In essence, the Turks were a heterogeneous society that ultimately was unified by the common denominators of language and religion.

SÜLEYMAN, THE GREATEST SULTAN

The Ottoman Empire reached its zenith during the reign of Süleyman (or Soliman or Suleiman) the Magnificent (1520–1566). This great leader extended the empire's borders from Austria to Mesopotamia and into North Africa. He also consolidated its internal strength. Unfortunately, he was also one of the last great leaders. After Süleyman's successes, the Ottoman Empire continued for some time to be a major player on the international scene, but never again would it attain such a level of strength. The last attempt to expand—to conquer Vienna, the capital of the Habsburg Monarchy—was during the War of 1683–1699, which ended in defeat and loss of territory for the Ottomans.

This photograph of Mustafa Kemal was taken in 1923, at the time that the independent Republic of Turkey was established. Kemal became known as Atatürk ("Father of Turks") and he is the most important figure in the history of modern Turkey.

between religion and state. To strengthen the newly independent country, he also relocated the capital deep in the interior. This maneuver was important both politically and economically.

PERIOD OF DEVELOPMENT AND POLITICAL UNREST

Atatürk died in 1938. During World War II, Turkey's government chose to remain neutral (not wanting to commit the same mistake the government had made during World War I when it sided with the Germans). Only in 1945, shortly before the end of the war, did Turkey formally declare war on Germany—a move that was designed to gain political benefits in the international arena. This gamble helped Turkey become one of the founding members of a new international organization, the United Nations. After the war, Turkey continued to develop both its democratic government and economy, yet it experienced many problems common to young democracies.

During the late 1950s, the country was experiencing growing political unrest and continuing animosity between the governing Democratic Party and the opposition. In an attempt to preserve Atatürk's principles, the Turkish military leaders staged a coup in 1960. After a short interim period and new elections, Turkey continued implementing democratic principles. The 1960s, during the height of the Cold War, brought turbulence and changes in domestic policy. In foreign policy, however, Turkey remained strongly pro-Western and pro-American, especially after becoming a member of NATO (the North Atlantic Treaty Organization) in 1952. The military again intervened in political life, both in 1971 and 1980, to ensure the continuation of reforms and to limit the influence of potentially dangerous Islamic fundamentalism. Today, conditions are similar in nature to those of previous decades. Political parties battle for dominance, while military leaders keep a watchful eye on the government and the country grapples with many social and economic problems. Despite the country's problems and the tremendous instability within the region, however, Turkey continues to progress.

CHAPTER

4

People
and Culture

This chapter provides a glimpse of the Turkish people and their way of life. An understanding of a people's culture—their language, religion, customs, ways of making a living, and so on—is essential. For example, without understanding the role of culture in the Middle Eastern countries, one will hardly be in a position to comprehend topics as far ranging as the impact of the Islamic faith on people's lives to rural and urban architecture. The ways in which humans adapt to, use, and change the natural environment—their cultural ecology—also give character to places. Human activity, for example, causes deforestation, desertification (the advance of desert conditions), the formation of artificial reservoirs behind dams, and countless other forms of environmental alteration. Turkey also has a rapidly growing population. The country's culture, ethnic groups, population, and patterns of settlement are the topics of this chapter.

ETHNIC GROUPS
Turks

Turks are the largest ethnic group in the country that bears their name, accounting for 80 percent of the population. Most of the world's ethnic Turks live in Turkey, although for reasons both historical and economic, several million reside in other countries. During the Ottoman period, Turks lived throughout the empire. Once the empire began to shrink in area, however, many people began to migrate into the territory of present-day Turkey. Migrations were particularly common during the several periods of military conflict in the twentieth century. During that time, millions of Turks migrated to Turkey from various countries of southeastern Europe. Today, countries such as Bulgaria, Macedonia, Serbia and Montenegro, and Ukraine still have substantial Turkic ethnic minorities. Some countries protected their minorities, including the Turks, whereas others did not. In the latter case, the dominant ethnic group sought to culturally assimilate their minority populations. Bulgarians of Turkic origins experienced the harshest treatment. Until the early 1990s, when major political changes occurred in the country, the Turks were not given independent ethnic status and were pressured to assimilate as ethnic Bulgarians.

In portions of Eastern Europe today, there exist strong xenophobic (fear of strangers and foreigners) attitudes toward Turks. This condition persists, at least in part, in response to animosities resulting from the centuries of occupation during the Ottoman Empire. Paradoxically, although xenophobia is quite widespread in this part of the world, hospitality toward foreign visitors and travelers is cordial. Turks themselves are not immune to xenophobia, because their country once contained a mixture of different ethnic groups. In 1915, thousands of Armenians were expelled or executed, as were many Greeks in 1922. Today, except for a minority of Kurds (who number about 20 percent of the country's population), Turkey is ethnically homogeneous.

In the assimilation process, religion was used to determine the ethnic status of citizens. This policy is a legacy of the Ottoman

Empire. Most of Turkey's citizens who belonged to the Islamic faith were automatically proclaimed to be ethnic Turks. For Kurds, an ethnic group distinctly different than Turks yet devoutly Muslim, this system of classification posed a problem. The government resolved the issue by proclaiming the Kurds to be Eastern Turks.

Outside Turkey's borders, Turks have many cousins related by ancestry and language. The largest groups in the Caucasus region are Azeris, who live mostly in Azerbaijan and Iran, whereas several others live in scattered locations on both sides of the Caucasus Mountains. Kazakhs, Uzbeks, Kyrgyz, Turkmens, and Uigurs live in the central Asian countries and western China. Domestically, Turks differentiate themselves by where they live, how they talk, and how their customs differ. Anatolian Turks, for example, have a different dialect than Turks living along the Aegean Sea or residents of the Istanbul area. Regionally, practices and traditions differ somewhat. This is especially true in the remote villages, where strong accents are still heard and folk (traditional) cultural practices persist.

Kurds

Kurds are an ethnic group that constitute a nationality (a people with a strong sense of self-identity) but are a people without a country. This maligned ethnic group is scattered throughout portions of at least seven countries within southwestern Asia. The majority of Kurds, however, live in Turkey. Actual population numbers are uncertain because of the political sensitivity of this ethnic demographic issue. Most estimates suggest that 15 to 20 million Kurds reside in Turkey today. For many years, the government banned any attempt to distinguish Kurds as a separate ethnic group. Fears of Kurdish separatism and the eventual creation of an independent Kurdish state prompted this seemingly harsh treatment. Rather than giving them autonomy, the Turkish government suppressed any attempt to establish a Kurdish national identity within Turkey's territory.

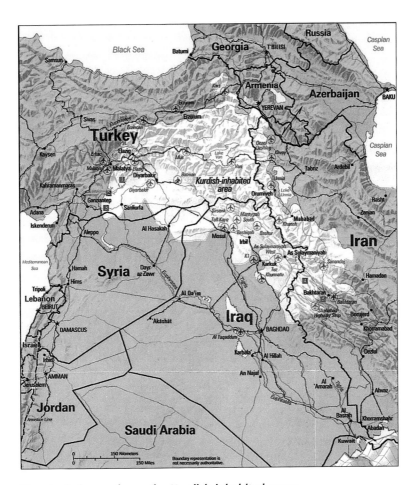

The shaded area shows the Kurdish-inhabited areas.

Any mention of the word Kurd or public use of the Kurdish language could land one in prison. Such treatment resulted in open hostility and rebellion in the 1980s. A precarious standoff exists today between the Kurds and the Turkish government, although there have been several recent signs of cooperation.

Kurds are among the most interesting ethnic groups in the Middle East, if for no reason other than their tremendous sense of independence. Despite numerous attempts by regional powers to assimilate them, Kurds always maintained a strong sense of self-identity. It is believed that Kurds were originally an

Indo-European tribe inhabiting central Asia and Persia. Eventually, they moved westward, where they live today. The southern border of Kurdish lands is in western Iran and northeastern Iraq. Hundreds of thousands of Kurdish people live in northeastern Syria as well. Kurds in Turkey live mostly in Eastern Anatolia in villages scattered throughout the countryside or in cities such as Diyarbakir, which is the unofficial Kurdish capital. In recent decades, large numbers of Kurds have left their homeland for urban centers in Turkey or Western European countries.

Arabs

Of the other ethnic groups residing in Turkey, only Arabs represent a significant numerical minority, with a population of about one million. Most Arabs live near the Syrian border, a region that did not become part of Turkey until the late 1930s. As is true of other non-Turkic citizens, Arabs feel a closer tie to their ethnic counterparts across the border in Syria than they do to the government in Ankara.

Others

At one time, several million Armenians and Greeks lived in Asia Minor. After World War I, the populations of these non-Muslim ethnic groups declined rapidly. Today, their combined numbers have shrunk drastically to fewer than 100,000.

Armenians were among the earliest residents of Asia Minor. Today, as a result of the Soviet Union's breakup in 1991, they have an independent country. Its boundaries, however, do not coincide with those of the historical Armenian homeland that was mostly on the Turkish side of the border around Lake Van. In 1915, after being accused of plotting a rebellion against Istanbul's government, most Armenians were deported from Anatolia. To this day, Armenians regard their deportation as an attempt at genocide. Turkey, of course, denies the accusations. Today, a very small number of Armenians live in Turkey's major cities, primarily Istanbul.

Turkey's Greek community is in a similar position. After the deportation of Greeks from Turkey and Turks from Greece in the 1920s, both countries "cleaned up" their problems with ethnic minorities. About 2 million Greeks were relocated into Greece and elsewhere. Today, only Istanbul and a few Aegean islands support small Greek communities.

Several small ethnic groups, primarily Muslims from the Republic of Georgia and Jews, constitute Turkey's remaining ethnic groups. In the 1990s, as a result of raging civil wars in the former Yugoslav republics, many thousands of refugees sought sanctuary in Turkey. Most were Bosnian Muslims. These people of Slavic origin accepted Islam during the period of the Ottoman Empire (1463–1878), and later in the twentieth century, on the basis of their religion, they were declared a separate ethnic group. Some of them returned home to Bosnia and Herzegovina after the conflicts ended, but many still live in Turkey.

RELIGION

Most Turks are Muslims, followers of the Islamic faith. Muslims, just as Christians, practice variations of the same religion. About 5 percent of Turks are Shiia Muslims, followers of the descendants of the prophet Muhammad's son-in-law, Ali. The great majority of people are Sunni Muslims. Because of Atatürk's reforms, Turkey is somewhat unique among Middle Eastern countries in that it is essentially a secular state. The government has worked vigorously to ensure the separation of church and state. Atatürk was not against religion. Rather, he believed that religion should not interfere with government. Religion, he felt, should be a personal matter, and every individual should be allowed to practice his or her own faith.

Considering the time and place of Atatürk's remarkable decision, he certainly was a visionary among world leaders. At the time of his reign, Turkey was mainly a rural country, where Islam played a primary role in all aspects of life, including government. Rural people living a traditional way of life

The Sultanahmet Mosque, or Blue Mosque, was built between 1609 and 1616 by the famous architect Mehmet. It is called the Blue Mosque because of thousands of blue tiles that decorate the interior. It is one of the largest mosques in historic Istanbul.

generally accept the status quo and are unwilling to welcome major changes. Atatürk, however, was a strong leader and had enough confidence in his leadership to revolutionize many basic elements of the prevailing lifestyle.

Today, religion is still important, yet its practice varies greatly between Turkey's urban and rural areas. In cities, where life is more dynamic, religion is less central to most people's lives than it is in the villages of the interior. Turkey allows the free practice of non-Islamic religions, especially the Orthodox Christianity of Armenians and Greeks. Istanbul, for example, is still the center of the Ecumenical Patriarchate of Constantinople. This legacy of Constantinople's past cosmopolitan glory retains the nominal leadership of Orthodox Christianity within the city.

LANGUAGE

The Turkic language, of the Ural-Altaic language group, underwent a transformation in the twentieth century. Although the Ottoman Empire was in many ways a multilingual society, Turkey's new government decided to purify the national tongue. It replaced Arabic, Persian, and other non-Turkic words in the vocabulary. Because Turkey had shrunk in land area and its boundaries no longer spread over parts of three continents, it was a good opportunity for linguistic reform. The Arabic alphabet was replaced with more modern Latin. Today, many young Turks also speak a foreign language, such as English or German. To a foreigner, Turkish is not an easy language to learn, mainly because of its complicated grammar structure.

POPULATION

Turkey has a population of approximately 71 million people, with a density of 238 people per square mile (616 per square kilometer). This figure, of course, is misleading. More than 10 million people are concentrated into the metropolitan area of Istanbul. In the relatively dry interior, however, densities in many places drop to just a few people per square mile. The population is increasing at a rate of about 1.5 percent each year. This figure is slightly above the world average of 1.2 percent but is well below western Asia's increase of 2.0 percent per year. Compared to most other European countries, Turkey's population is quite young: 28 percent of its people are under 15 years of age.

Higher education is somewhat different than in the West. This is true primarily because of government controls that contribute to a lack of autonomy in the decision-making process. Governments of many countries, including Turkey, believe that institutions of higher learning serve as a breeding ground for antigovernment feelings. For that reason, the government appoints university presidents and distributes funding. This practice cripples the university system in many ways.

Despite the problem of government intervention in university affairs, Turkey has dramatically improved its system of higher education. Today, more than a half-million students are working toward degrees in dozens of different institutions of higher learning. Considering the size of the Turkish population, the number of university students seems relatively small, especially compared to the United States. The difference lies in differing ideological approaches. In the United States, the general policy is that "no child should be left behind." State and federal governments work to provide an opportunity for anyone who wants it to pursue a higher education. In Turkey, the approach is different. The country follows much more closely the typical European model. Pursuit of higher education is an honor reserved only for the best and brightest. Acceptance is also highly competitive. A rigorous exam is taken by all high school graduates seeking to continue their education. The number of openings in both universities and individual fields of study are limited. Success on the exams, rather than individual university policies and preferences, determine whether or not one is accepted.

CONTEMPORARY LIFESTYLE

The vast size and cultural influence of the Ottoman Empire left numerous imprints on the lifestyle of people throughout southeastern portions of Europe as well as the Middle East. As a youngster growing up in the region, the author was unaware of the tremendous cultural legacy resulting from centuries of

The life of the rural people is entrenched in traditional practices. The people shown live in a village called Asagi Konak, which is near the Iraqi border in the extreme eastern part of Turkey.

interaction with Turks. Everyday activities in the rural areas of what is colloquially known as the Balkans (Europe south of the Danube and Sava rivers), however, resemble in many ways those of the century before, or earlier. Architecture and many customs in this region are more Asian than Western European. A village in Bulgaria or Macedonia is apt to have more in common with a village in Anatolia than with its counterpart in other Slavic countries such as Poland or the Czech Republic. Exploring the lifestyle of Turks brings to our attention the importance of cultural interaction between different groups of people inhabiting the same living space. Contemporary Turks, on the other hand, have accepted various cultural traits from the West. This is particularly evident in urban centers. There, Western influence not only came through Atatürk's political decisions to modernize the country but also through cultural diffusion (spread) from elsewhere. Such conditions have essentially created two Turkeys: One is the folk culture of rural people, whose way of life is deeply entrenched in traditional practices; the other is the fast-paced, ever-changing popular culture of the growing urban population.

A traveler through Anatolia will surely notice many small groups of tents scattered around the countryside, surrounded by domestic animals, such as goats and camels. These people still live as nomads, moving from place to place in the rugged terrain of Turkey's interior. Centuries ago, many people of Asia Minor lived a nomadic life. They constantly migrated, carrying their few meager belongings, following the seasons, and searching for better pasture conditions for their flocks. Their animals provide the main source of food, building material, mobility, and capital (animals and animal products could be sold or exchanged for other necessities). It should not be surprising that the nomadic people still exist in many countries. The lifestyle seems hard compared to sedentary life in villages or towns, but nomadism is another example of deeply engrained cultural practices, with origins dating back hundreds or even thousands of years. Turks, themselves a nomadic people from central Asia, supported this type of lifestyle in the early days of their reign in the Anatolian plateaus and hills. Interestingly enough, many foreigners find it hard to comprehend the existence of American trailer parks, which certainly have some similarities to the nomadic experience. If a particular place is not suitable for some reason, simply take your house and all other belongings and relocate wherever you like!

The life of sedentary people in villages is slowly changing. Some settlements are continuations of millennia-old communities established at times when the region was predominantly Christian. Generations of villagers continue to follow a lifestyle closely resembling that of their ancestors. Various agricultural activities and small shops are the main economic activities. In

The valleys of Cappadocia are located south of Erciyes, a formerly active volcano. Some of these chimney-like rock formations have been dug out to provide living quarters. A variety of historic monastic ruins, like these anchorite dwellings, are found in Goreme National Park. Anchorites were people who lived in seclusion for religious reasons. Some modern-day Turks continue to live in rock dwellings in this area.

recent decades, however, places such as Cappadocia discovered the lure of tourism and the income it can bring. Outsiders, they learned, are willing to spend money to visit scenic natural and cultural attractions and a well-preserved past. Cappadocian rock-based settlements are one of the main attractions to people visiting Turkey. Thousands of years ago, people realized that carving into certain types of volcanic rocks was a relatively easy and rewarding process. Instead of building settlements that are hard to defend and provide little protection from the weather, the ancient inhabitants of Cappadocia carved whole cities in rocks. The practice may have begun with early Christians searching for shelters in times of persecution. Others continued the practice. At one point, some of these underground cities had between 10,000 and 20,000 residents. Contemporary Turks still live in the rock-based dwellings, which provide comfortable and easily maintained living spaces.

Contrary to the idyllic countryside, city life bustles with a frenzy of activity. The larger the urban center, the faster paced and busy are the lifestyles. This is particularly obvious in Istanbul, one of the world's great cities. Here, everyone seems to be perpetually "on the go." Tourists to the city are often shocked at the locals' amazing ability to expand a two-lane road into four lanes of chaotic traffic! Crowdedness of certain areas is usually related to economic activities. Few tourists ever visit Istanbul or other Turkic cities without going to a *bazaar*, the original version of modern malls. Bazaars are strung-out markets where one can find anything imaginable—from "ancient" antiques, jewelry, and carpets to food of any kind, souvenirs, and affordable leather products. Istanbul's most famous bazaar has hundreds of different small-sized stores with thousands of products displayed.

There are only two rules that a shopper must know before exploring the bazaars. First, be ready to bargain for the best price; second, be ready to pay once you have agreed on the final price. Bailing out of an agreement is considered to be a dishonest act in a society where personal pride and honor are held in high esteem. Many Westerners enjoy bargaining even for the cheapest products, believing that this is perhaps the best way to experience local culture. Shopping for a carpet is perhaps more interesting than shopping for any other product. Their variety is endless. Turks are known for manufacturing wonderful oriental-type carpets. Shops often carry carpets from other Asian countries, from Iran to Kazakhstan, as well. To an experienced shopper, each carpet tells a story about its region and culture of origin based on the design and weave. Carpet makers in Iran use different knots, for example, whereas each tribe or clan from central Asia uses different patterns and

Bazaars are lines of stalls that may sell anything imaginable. The spice bazaar, or Egyptian bazaar, is located in the old section of Istanbul.

colors. Oriental rugs are often rather expensive. With successful bargaining, a shopper can lower the original price hundreds of dollars.

Turks are passionate people who appreciate the good life. Restaurants or various smaller coffee and dining shops are the places of entertainment where locals congregate to visit with friends and family. The atmosphere is always lively and exciting, with conversations ranging from domestic and foreign politics to the achievements of favorite sports teams, especially football (soccer) clubs. In Istanbul, comments on major football clubs such as Galatasaray, Besiktas, or Fenerbahce can make numerous new friends for a foreign visitor. That, however, is only in the case when such comments are directed to the right fans. Football (soccer) in Europe is elevated to the position of a "religion." Dining out is not a quick in-and-out experience; rather, it is a remarkable, often hours-long, event that includes lots of dancing and singing. Another good form of relaxation and entertainment is to visit one of the famous Turkic baths. These facilities offer bathing, massages, or simply soaking in the water to relax after a hard day's work. Turks regard visits to baths (locally known as *hamams*) as an occasion for social interaction among friends. This essential element of Turkic culture is practiced today throughout much of the world, having been spread primarily by colonial British.

CUISINE

Diet, according to many geographers and anthropologists, may be the best cultural indicator of a particular society. Two centuries ago, famous French gourmet Anthelme Brillat-Savarin (1755–1826) claimed that people's identities are the reflection of their diets by boasting, "tell me what you eat and I'll tell you what you are." Turkish cuisine, like Turkish society, reveals excitement and simplicity. Dining is a collective experience that begins with soups, sliced meat, and perhaps vegetables as hors d'oeuvres known as *meze*. The meal continues with the selection of a main course and concludes with delightful desserts and coffee. For many Turks (despite their Islamic faith, which prohibits the consumption of alcohol), meals also have a good companion in alcoholic drinks, particularly raki.

Contemporary Turkic cuisine is a blend of influences from both the Middle East and European provinces of the former Ottoman Empire. In many ways, it is a Mediterranean cuisine that features fresh ingredients. The preceding paragraph, in fact, could have been written about Bosnian cuisine with little, if any, change. Raki, rakija, or prakija, for example, is a popular drink consumed throughout southeastern Europe. It is a form of brandy and can be distilled from plums, apples, pears, or any other fruit. Researchers believe that distillation of alcohol first began in Asia Minor and later spread to the rest of Europe and the world.

Some Turkish cuisine has been accepted globally. Kebab and baklava are two of the best-known foods to have gained widespread acceptance. Although many Middle Eastern countries might claim these culinary delights as their own, the author prefers to identify them with Turkey. Kebabs are pieces of lamb meat trimmed from the bone and prepared in different ways, from grilling and broiling to stewing. It can be prepared in hundreds of different ways, because each town has its own special recipe. One of the favorite ways to prepare a kebab is to grill the meat on skewers over smoky charcoal—a shish kebab. Baklava can be found relatively easily in the United States, either already prepared or the ingredients ready to combine into this flaky pastry combination of honey and nuts.

Turkish desserts in general tend to be very sweet. They offer good company to strong Turkish-style coffee prepared in small copper pots. Families often have beautifully decorated copper pots that are used only for special guests. Coffee is served in tiny tulip-shaped cups with, depending on individual taste, sugar or milk. Turkish coffee is unique for its unfiltered thickness, which gives it a strong, yet delicious taste. Street vendors and small restaurants always offer coffee and sweets as quick refreshments. Of all Turkish products, the author's favorite is borek, which is a form of rolled pastry served baked, boiled, or fried. Easy to make yet hard to perfect, borek is a quick, popular snack that can be either sweet or salty, depending on fillings. The most popular fillings include minced meat, cheeses, potatoes, and spinach, yet any other available ingredient (such as pumpkin) tastes good as well.

CHAPTER

5

Government

V ia military interventions and attempts to reform political and economic life, Turkey has been slowly developing democracy and building democratic institutions. Few countries have experienced a faster or more radical transformation into a Western-style democracy in a shorter period of time. Attempts to introduce administrative reforms, however, can lead to political instability. This is particularly true if reforms are a step ahead of what a society is willing and able to accept. In some respects, this is what happened to Turkey since its independence in 1923. Today, however, Turkish society willingly accepts leaders from all sides of the political spectrum. Even those labeled "pro-Islamic" successfully participate in the political process.

SEPARATION OF POWER

The separation of power, as in most other parliamentary democracies, is divided among the presidency, the National Assembly, and the judiciary. Executive powers are in the hand of the president and the Council of Ministers. In most Western European democracies (except France), the president serves primarily a ceremonial role and possesses limited executive power. In Turkey this is not the case. Culturally, Turks (and their regional neighbors) appreciate and desire a strong leader; therefore, presidential powers regulate many elements of the nation's political life. The president's power increased significantly after military interventions that led to a revision in the constitution to expand presidential powers.

The president's duties include, among others, recommending and appointing a prime minister from among the elected members of the National Assembly, yet presidential powers reach much farther. For example, he or she holds the power of veto on any legislation and over appointment of other officials as well as other forms of intervention that make him or her one of the strongest players in Turkey's political life. Presidents are limited to a single seven-year term. Of the three branches of government, the executive branch appears to be the strongest. Its strength is based on the country's most recent constitution, which was written and adopted after a period of military intervention in government affairs during the 1980s. The prime minister and the Council of Ministers—a cabinet—in traditional practice are more answerable to the president than to the national legislative body.

The Turkish Parliament has 550 members elected by the principle of proportional representation, a system popular in Europe but not in the United States. It is a rather complex electoral system that limits the polarization between the two largest parties in parliament. The system allows the participation of smaller players, if their party receives at least 10 percent of the vote in the national elections. Members of parliament, representing the electoral district of Turkey's 81 provinces, serve four-year terms. Every Turkish citizen older than 18 years of age is allowed to vote, and voting is considered a duty of each citizen. The Turkic constitution also requires another duty, serving in the military forces. Every male is required to serve in the armed forces on reaching 25 years of age.

Several courts with different duties form the judiciary branch. These courts are supervised by the High Council of Judges and Prosecutors, a body of experts appointed through presidential decisions. The Constitutional Court considers cases involving problems with the legislation. It and the Court of Appeals form the two supreme courts of Turkey's judicial branch. Other courts are the Council of State (administrative issues), the Court of Accounts, and the Military Court of Appeals.

POLITICAL ISSUES

Various domestic and foreign issues are permanently present in Turkey's political life. Today, the most important concerns are the preservation of a secular state; allowing, yet limiting the influence of, political parties with strong religious agendas; the always prickly Kurdish issue; the search for a political solution to the conflict in the ethnically divided island of Cyprus; membership in the European Union; and the always-present economic difficulties. The U.S. intervention in Iraq was one of the burning issues in late 2002 and early 2003. The Parliament had to consider whether to allow the transport of American troops from or over its territory.

The most important long-term political issue is the continuing search for a solution to the question of Kurdish nationalism. After years of armed rebellion against Ankara's government and numerous political deadlocks, Kurds are finally achieving certain long-fought-for freedoms. Turkey maintains an extensive military presence in areas with a strong Kurdish majority, yet clashes between the army and Kurdistan Workers' Party (PKK) have slowed down recently. Kurds began an armed resistance to the Turkic regime in 1984. Hostilities continued throughout the 1980s and 1990s under the leadership of Abdullah Ocalan. Kurds saw Ocalan as a charismatic figure who would lead them toward independence. The government in Ankara, however, regarded him as a terrorist and murderer responsible for more

These Kurdish women and their children live in a shanty town just outside the wall of the mainly Kurdish city of Diyarbakir.

than 35,000 deaths in the 15-year-long rebellion. Ultimately, Ocalan was captured in Nairobi, Kenya, and sentenced to death in 1999. The government, however, decided not to elevate him to martyr status by execution; rather, he is now serving a life sentence in prison.

The Turkish-Kurdish conflict is a rather complicated issue in which both sides accuse the other of terrorism and atrocities. Turkey justifies its actions as a legal attempt to preserve its territorial integrity against Kurdish Marxist-Leninist forces. Kurds, on the other hand, justify their actions as being the only way to preserve their national identity. Turkey, you must remember, never wanted to recognize the Kurds as a separate ethnic group. Many constitutional changes created additional animosity. These changes include bans on many Kurd-related civil rights issues, ranging from freedom of speech to education in the Kurdish language.

Turkey's domestic problems have also become a topic of international debate. Human rights organizations and some governments have accused Turkey of implementing policies that resemble a dictatorship rather than a modern democracy. With Turkey desperately seeking membership in the European Union, these accusations hurt. As this book goes to press, Turkey has the first single-party government in nearly two decades. Prime Minister Recep Tayyip Erdogan, who was a popular mayor of Istanbul before leading the Justice and Development Party (AK), faces many challenges. None, perhaps, is more important than making changes that will clear a path for Turkey's future membership in the European Union.

Another government priority is finding a solution to political problems on the island of Cyprus. There, Turkic and Greek Cypriots have lived separated for the past three decades, under the vigilance of United Nations' peacekeeping forces. Long-lasting animosities between Turks and Greeks (not only on Cyprus, but elsewhere, as described in Chapter 3) reached a climax in 1974, when Turkey's forces invaded the island to protect Cyprus from falling into Greece's hands. Greeks, on the other hand, saw the invasion as an excuse for occupation. International intervention divided the island, leaving ethnic Turks on the northern end of Cyprus and their Greek counterparts in the south. The line of demarcation sometimes divides cities in half, as in the former capital of Nicosia. In the years after Turkey's invasion of the island, Ankara pushed for migration from the mainland to Cyprus, hoping to increase the percentage of Turkic Cypriots on the island. Greek Cypriots rejected this act as a further sign of occupation and continued their own development along the southern coast.

Contemporary Cyprus has two faces: the economically prosperous Greek side and the declining and desperate Turkish

Talks between Turkish and Greek leaders over the reunification of Cyprus took place at the United Nations on October 4, 2002. Rauf Denktash, Turkish Cypriot leader (left), and Greek president of Cyprus, Glafcos Clerides (second from left), met with the UN Secretary General, Kofi Annan.

side. Both parties continuously claim their willingness to find a suitable solution leading to the reunification of Cyprus. Interestingly, the European Union (EU; of which Greece is a member) announced that it will accept the Greek-controlled part of Cyprus to EU membership in 2004 if a solution has not been found. Because of this threat, Turkey has become actively engaged discussions leading to possible reunification. There are problems, however. Powerful Turkish military leaders are unlikely to support reunification because of the strategic importance of Cyprus for defense. Membership in the European Union would be of enormous political and economic benefit to Turkey. The country applied for membership in 1987 and has been on the waiting list longer than any other country, however. Considering the many domestic and international conflicts the country faces, its waiting period may be extended many more years. The EU issue places the government in a "catch-22" situation. National pride, which is highly regarded by Turks, must be sacrificed for future economic and political well-being. The alternative is for Turkey to become increasingly isolated from Europe and its best hope for development.

Economic issues can be the important problem facing any government. Governments, therefore, work intensely to improve the national economy in order to encourage their future political gains. Governments of poor, developing countries, of course, usually face more difficulties than do those of developed countries. There is never enough cash to fulfill demands or to expand existing or implement new programs. As a developing country, Turkey's government faces critical financial problems. Since the beginning of the Turkish Republic in 1923, economic issues have often been catalysts of political change. This certainly has been true during recent years. For example, Turkey has borrowed heavily from the International Monetary Fund, the World Bank, and other organizations in order to further develop its economy. This borrowing, however, has left the country national and international debts of more than $100 billion. The government also faces problems of steadily rising inflation and slowing economic growth.

As a member of NATO, Turkey has a good military relationship with Western powers, including the United States. During the Gulf War in 1991, military bases in Turkey were used for action against Iraq. In 2003, however, the National Assembly voted against the U.S. use of military bases in Turkey, perhaps because of public opinion. In this conflict, nearly 90 percent of Turks supported a political, rather than military, solution to the problem. Their reason was not necessarily "anti-American." Rather, many feared that the dissolution of Iraq would build a foundation for the creation of an independent Kurdish state. Were this to happen, it would pose a direct threat to Turkey's national interests. Use of Turkish bases was denied despite the fact that the United States was prepared to support Turkey with $30 billion in grants and loans in return for its support. This offers a splendid example of the way in which politics, economics, and national interests are integrated.

CHAPTER

6

Economy

T urkey faces many economic challenges. As mentioned in the previous chapter, in order to stimulate growth and restructure the economy, the country had to rely on borrowing from international financial institutions. Times have long passed when independent economies could exist without major contact with the rest of the world. Successful integration with the global economic community is a primary determinant of economic success, even for the most isolated countries. Global changes strongly influence any economy, including Turkey's. In the 1990s, Turkey's gross domestic product (GDP) was rising steadily. It was on the basis of this economic strength that the country was able to secure billions of dollars in grants and loans. The money, of course, was intended to support even greater economic development in the country.

Unfortunately for the Turks and many other countries, during recent years the global economy has slowed. Turkey became a victim of such conditions. The country's economic growth was dramatically reduced, leaving billions of dollars in unpaid loans. Currently, the country's foreign debt is about $120 billion. This is an enormous amount for a country that is battling shrinkage in its GDP (GDP is around $470 billion) and is facing inflation that is rising toward 50 percent per year. Still another problem, one common to many developing countries, is rampant corruption. This condition, for which Turkey's leaders are often criticized, discourages desperately needed foreign investment.

During the early years of the twenty-first century, Turkey is suffering from a serious recession. Only fiscal responsibility, political stability, and an end to corruption will reverse the current economic malaise. This, however, will take time to accomplish. This chapter presents a brief overview of Turkey's economic potential and realities.

AGRICULTURE

Agriculture is crucial to Turkey's development. Its primary importance rests less in its contribution to the country's sluggish financial system than that to the country's socioeconomic well-being. One-third of Turkey's males and two-thirds of the employed females work in the agricultural sector. This may seem paradoxical because for years the government has placed emphasis on industrial development, leaving agriculture in a continuous state of decline. Just 50 years ago, agriculture accounted for half of Turkey's GDP; today, it contributes less than 15 percent. With so many people dependent upon agriculture for their livelihood, however, the transition from agriculture to industry as the country's economic base will be difficult.

Turkey is blessed with land on which two high-quality export crops—tobacco and cotton—can be grown. For decades, Western cigarette manufacturers have been using the Turkic-grown tobacco for their best brands. Although the use of tobacco is declining in the United States, consumption is growing worldwide. This is bad news

These are tobacco leaves drying in Amasya. Tobacco is a high-quality export crop for Turkey, which ranks fifth in the world for tobacco production. Turkish tobacco has been used in the manufacture of Western cigarettes for decades.

in terms of international health, but it is certainly good news for growers of prized Turkic tobacco.

As a predominantly Muslim country, the consumption of wine is limited. Nonetheless, Turkey ranks fifth in the world in grape production, right behind the other heavyweights Spain,

France, Italy, and the United States. Most grapes grown in Turkey are destined to become ingredients in food, but a careful restructuring of the crop toward wine production could rejuvenate the agricultural sector. Cotton continues to be the primary export crop. Other export crops include olives and olive oil, a number of fruits and vegetables, and sugar beets. Some livestock also are exported.

ENERGY

In terms of energy, Turkey consumes more than it produces, leaving the country with an energy deficit. This is ironic, because the country lies adjacent to the world's largest reserves of petroleum and natural gas! Geologic structure did not bless Asia Minor with larger amounts of natural gas or oil. Even Turkey's coal deposits are sparse and of lower quality. The lack of adequate energy resources hampers the country's plans for economic development. Turkey must rely heavily on costly imports that further drain its fiscal resources. In 1991, the United Nations placed an embargo on Iraq's oil production and export. Although directed toward a belligerent Iraq, the embargo dealt yet another blow to the Turkish energy sector. Because of its favorable geographic location, Turkey had become a transit country for Iraqi petroleum exports. Before the first Gulf War (1990–1991), oil from Iraq's northern field around Kirkuk and Mosul was being shipped to the port of Ceyhan. It is the largest port in Turkey and one of the larger in the Mediterranean Sea. Ceyhan's port facility can serve the largest tankers. This is exceptionally important for transport to North America, because it drastically reduces transportation costs. For more than a decade, Turkey was deprived of royalties from the transport of Iraqi oil. Sanctions were lifted in 2003 (after the second Gulf War), and transport will resume. However, it may take Iraq 5 to 10 years to reach pre-1991 levels of production.

Turkey has struck another significant deal for the transport of oil. After many years of negotiations, an agreement was finalized to bring Turkey closer to central Asia's enormous energy reserves. A pipeline will transport petroleum from Azerbaijan's capital, Baku, to the port at Ceyhan. In selecting this route, geopolitical circumstances favored Turkey. Some governments, including the United States, supported oil transit over Turkic, rather than Russian or Iranian, territory.

Rather than exploiting small and expensive domestic fields, Turkey intends to meet its demands with energy imports. Because of the recent economic recession, energy consumption has gradually decreased. Once the national economy recovers, however, demand will increase. At that time, Turkey must obtain enough energy to expand its industrial capacity. Most observers believe that, for both economic and environmental reasons, the worldwide consumption of natural gas will increase greatly in the near future. Realizing that, Turkey cooperated with Russia—the country with the largest reserves in the world—in building the pipeline across the bottom of the Black Sea. The line connects the two countries and can be extended toward the Mediterranean.

One energy source that Turkey does have in abundance is hydroelectric power. Currently, one-fourth of the country's electrical energy comes from dams (the rest comes from fossil fuels). Though building large dams is often criticized by environmentalists concerned about the dams' environmental and human impacts, generating energy in such a way is more acceptable than building nuclear power plants. Major hydroelectric projects are operated in Southeastern Anatolia, mainly on the Euphrates and Tigris rivers and their tributaries. Currently, there are dozens of power plants, and the country plans to build even more. The largest facility is Atatürk Dam, located on the Euphrates, with the capacity to generate 2,400 megawatts of electricity.

INDUSTRY AND SERVICE

Half of Turkey's work force is employed in the rapidly developing service sector. Although expanding, Turkey has not yet reached the status of a postindustrial society. Productivity still remains lower than in the West. Poor countries face many problems, including weak governments and state control over businesses, as they attempt to develop. In Turkey, these conditions have improved during the past several decades. There have been major advances, for example, in foreign investments, including banking, in the country's economy.

Turkey's beautiful physical landscape, hospitality, and affordable prices (especially when compared with European countries using the euro currency) have begun to attract millions of tourists each year. Tourism now adds more than $10 billion of much needed foreign currency to the country's economy annually. Numbers continued to increase steadily until 2003. Because of the war in neighboring Iraq, however, tens of thousands of potential visitors decided to cancel reservations and change destinations. They did not want to risk being too close to the combat zone. Turkey's tourism potential is boundless. With a well-coordinated program to advertise, promote, and facilitate tourism, the country could become one of the world's major travel destinations. It offers a splendid variety of attractions—from towering mountains to shining seas and a rich history to fascinating culture. Istanbul itself is not only a huge city but is one of the world's most exotic.

Industrial production accounts for one-third of Turkey's gross domestic product and about one-fifth of all employment. The country has diversified the manufacturing sector of its economy more successfully than have many developing countries. Rather than depending on one or two primary industries, Turkey produces many things. Its major goods include textiles, various chemicals, agricultural products, minerals, assembly of automobiles, and

Turkey's beautiful physical landscape, such as this locale on the Turquoise Coast, has begun to attract millions of tourists each year, but the war in Iraq has dampened the enthusiasm of many potential visitors.

consumer electronics. In recent decades, textiles have been the leading export product.

LINKAGES

The current recession has further slowed development of Turkey's transportation infrastructure. One of the government's chief tasks is to connect remote underdeveloped areas with the rest of Turkey. After World War II (in 1945), attention focused mainly on expanding the roadways. The road network grew to the current 237,374 miles (382,016 kilometers), although 171,000 miles (275,198 kilometers) remain unpaved. In the near future, Turkey plans to build 9,300 miles (14,967 kilometers) of new highways that will connect major cities. It also plans to improve the existing networks within urban areas that have long outgrown their traffic capacity. During the second half of the twentieth century, very little work was done to expand or improve upon the railroads. Currently, the country has 5,344 miles (8,600 kilometers) of rail lines, 1,305 miles (2,100 kilometers) of which are electrified. Turkey has 120 airports currently in use, but only 86 have paved runways.

Although the number of airports continues to grow, flying plays a secondary role in domestic transportation because of its high cost. Aviation provides a crucial link between Turkey and the rest of the world, however. Turkish Airlines, the national carrier, has dozens of regularly scheduled daily flights to many international destinations. Its chief hub, Istanbul, is also served by a number of other international carriers. Most cities of any size have regularly scheduled air service. During the tourist season, charter flights also serve many smaller markets.

Commercial shipping has long been neglected. Only during the past several decades has the government finally attempted to stimulate its development. It seems strange that a country surrounded on three sides by water plays such a minor role in international shipping. The answer lies in Turkey's historical development. The Ottoman Empire was a land-based power. The Turks themselves were of nomadic origin. Their home was the interior—the vast steppes of central Asia and the semidesert regions of the Middle East. Turks turned to the sea only when they had to. Their neighbors, the Greeks, on the other hand, always lived by the sea and from the sea. It is not surprising that their trading activities included shipping. The richest and most powerful shipping mogul of the twentieth century, the Greek tycoon Aristotle Onassis, was born in present-day Turkey.

Turkey is still a step behind the West with regard to developed communication networks. The number of phone lines, television and radio stations, and Internet users is lower (per capita) in Turkey than in the rest of the European Union. This is understandable, considering Turkey's late start toward modernization. The country's size, widely scattered population, and often rugged terrain also posed obstacles. Rural areas in the interior were hard to reach and expensive to connect. Compared to what was in use before the communication network was modernized, Turkey has done remarkably well. It has been able to bypass some conventional means of communication (e.g., standard phones) and build modern networks such as cellular phones and Internet providers. The country boasts having as many cell phone users as users of traditional phones (about 20 million each). There are also millions of Internet users, and their numbers increase significantly each year. More than 600 television stations operate throughout the country. Some stations broadcast their program over satellites to Germany and other European countries that host thousands of Turkish-speaking residents.

TRADE

Attempts to modernize Turkey have had a negative affect on the country's trade balance. Even though the import-export ratio has improved somewhat during recent years, the value of imports still exceeds that of imports. During the early years of the twenty-first century, imports averaged around $40 billion, whereas exports were valued at about $35 billion. Inflation and the weak state of national currency, the Turkish lira, contribute to the negative balance. To buy products on the international market, Turkish companies must present hard currency, such as the dollar or euro. At the same time, with the lira becoming weaker, the hard currency becomes more costly to obtain. Trading partners are mainly European Union countries, the United States, and Russia. With the exception of Russian trade, the ratio between imports and exports is about the same. There is in imbalance with Russia because Turkey imports much more from than it exports to that country. Most of the difference results from Turkey's heavy dependence upon Russia as a source of fossil fuels.

7

Cities
and Regions

R egions—unless they are delineated by political adminis-
trations—are geographers' creations. The Middle East, for
example, may include only countries of Southwest Asia to
one geographer, whereas another might extend the region well into
North Africa. Regions are geographers' "convenience packages." The
concept is used to make geographic study easier by recognizing areas
of the Earth's surface that are similar in some way(s). Turkey can be
divided into a number of different regions, each of which differs
somewhat from all others. In some instances, location or characteristics of the physical landscape set the region apart; in others,
aspects of history, ethnicity, or economic activity make the region
unique. For this book, the author, like most Turkish geographers, has
divided Turkey into seven major regions. They are Marmara, Black
Sea, Aegean, Mediterranean, Central Anatolia, Eastern Anatolia, and
Southeastern Anatolia.

92

The Byzantine city of Constantinople was renamed Istanbul after the fall of the city to the Muslim conqueror Mehmet II. Today, the Galata Bridge crosses the Golden Horn at the city and port of Istanbul. The Golden Horn is considered to be one of the best natural harbors in the world and both the Ottoman and Byzantine navies and shipping interests were centered here.

MARMARA

Marmara is the smallest, yet the best developed, of Turkey's regions. It is the country's economic and population "heart." The region covers the country's extreme northwest, including European Turkey and area bordering the Sea of Marmara. With the Bosporus and Dardanelles—the straits linking the Sea of Marmara with the Black Sea and Aegean Sea, respectively—Marmara was historically predestined to become Turkey's leading region. Istanbul's early rise as a major urban, industrial, transportation, and educational center was crucial for the development of other smaller cities in this region. They benefited tremendously from their proximity to the country's largest and most influential city.

As is the case in many developing countries, capital and investments are often unequally distributed. Most capital resources go to benefit urban populations, leaving little for rural development. Istanbul consumes more than any other Turkish city, but it also generates one-third of the country's total gross domestic product (GDP). Istanbul has been an important transportation center—one of the world's leading "crossroads" locations—throughout its history. With the advent of railroads, the city was the final destination of the famous Orient Express rail service that used to connect Paris and Istanbul. It was also the beginning of the Baghdad Railway, which linked the city with Baghdad, in present-day Iraq. Istanbul is also the heart of the Turkic financial market. It is home to some 50 percent of Turkey's companies and a thriving port through which much of the country's foreign trade is conducted.

Istanbul's population is approaching 15 million, placing almost one of every six of the country's citizens in the city. During the last half-century, the city's population exploded, doubling several times. Rapid growth has continued. Between 1990 and 2000, the city grew by 2,200,000, and its growth continues at an annual rate of about two percent. It is impossible for Istanbul to keep pace with providing adequate services for the booming population. The government, of course, would like to see the trend reversed. It encourages people to move to other regions of the country, where population growth is needed to fuel economic development. Other major cities in Marmara are Bursa (1,200,000), the Ottoman Empire's first capital, and Izmit (196,000), a city with growing industrial production.

BLACK SEA REGION

The Black Sea region extends from Marmara on the west to the Republic of Georgia border on the east. Its southern border follows the ridges of the Pontus Mountains; this influences population distribution and density in a drastic way. Settlement is clustered close to the Black Sea on the narrow coastal plain.

Because of its mild climate and fertile alluvial soils, the coastal plain has supported a high population since ancient times. Highest population densities occur along a 300-mile (483-kilometer) stretch between the Kelkit River and the Georgian border. Here, as many as 300 people are crowded into each square mile (115 per square kilometer).

Successful commercial agriculture dominates the economic activity of the eastern part of this region. In the west, the emphasis is on industry, particularly the mining of coal and production of steel. Zonguldak and Karabük are important centers of Turkish heavy industry. Although the region has several major urban centers, much of the population resides in scattered rural settlements. Major metropolitan centers in the region include Samsun (363,000), which is the largest city in the Black Sea region, followed by Trabzon (215,000), Ordu (113,000), Zonguldak (104,000), and Karabük (101,000). Except for Karabuk, an iron and steel producing industrial center located inland near the mountains, all these cities are coastal settlements. Samsun is Turkey's most important port on the Black Sea. Because of the area's productive commercial agriculture and its import-export trade, the city is growing rapidly. In fact, its population grew by 30 percent between 1990 and 2000. Trabzon is one of Turkey's oldest cities. Greek colonists founded it in the eighth century B.C. For centuries, the city has played an important role as a regional trade center. Today, Trabzon is the major port for the export of agricultural products grown in northeastern Turkey.

AEGEAN REGION

The Aegean region—western Turkey bordering the Aegean Sea—has long played a vital role in the history of Asia Minor. Its location gave it a window not only on the Aegean and Mediterranean seas but also on Greece. Although the region was once controlled and populated mainly by Greeks, today few Greeks remain. The region has played a vital role in the development of the Turkish state and its economy.

Great scenic beauty and a fascinating history make this one of Turkey's most promising tourist destinations. The area boasts many archaeological sites dating from the classical period, including the famous Troy. In addition, the Mediterranean climate and beautiful coastline and scattered islands have contributed to recent rapid development of the tourist industry. The Aegean region's land area covers the drainage systems of the Buyuk Menderes and Gediz rivers. In the interior, settlement tends to cluster as a narrow ribbon in the fertile river valleys. Most of the region's residents, however, cling to the coast. The interior area bordering Central Anatolia is sparsely populated. This pattern is similar to that of other regions in Turkey and elsewhere that have access to the sea. Although the Aegean Sea plays a major role in the region's economy, Turkey controls only a few of the islands near its coast. Because of twentieth-century geopolitical decisions, almost all of the Aegean islands belong to Greece. A well-developed network of ferries connects Turkey with the Greek islands and mainland Greece as well.

The region's economic, transportation, and educational center is Izmir (2,232,000), Turkey's third largest city. As is true of other urban areas, Izmir has experienced explosive population growth during recent decades. For several millennia, the ancient city was known by its Greek name, Smyrna. After the Turks claimed the region, however, its name was changed to Izmir. The city has a strong local manufacturing base. It is also Turkey's second seaport (after Istanbul), exporting a variety of agricultural and industrial products.

MEDITERRANEAN REGION

The Mediterranean is similar to the region bordering the Black Sea in that most settlement is concentrated in the area of coastal lowlands. Inland, the towering Taurus and Anti-Taurus Mountains dominate the background and form the region's northern border. As elevation increases, population density sharply decreases. Geographically, the Mediterranean region extends along the Mediterranean Sea from the vicinity of Rhodes Island (Greece) in the west and continues eastward to the Syrian border. The coastline here is known as the "Turkish Riviera" and, as the name implies, is another popular tourist destination. Because of the deep blue-green beauty of the seawater, it is also called the "Turquoise Coast."

The region is dotted with ancient settlements, many of which are modern cities today. These cities attest to the fact that Turkey's Mediterranean region, particularly the southern coast, has been a popular place to live for many centuries. Most of the larger cities are located in the eastern lowlands on the plains surrounding the cities of Adana and Antalya (Antakya). Settlement along the coastline is limited to preserve the picturesque atmosphere of smaller communities.

The region has a number of urban centers, including Icel, Tarsus, Osmaniye, and Iskenderun, which are clustered in the eastern part of the Mediterranean region. They, and the region's largest city, Adana, occupy the sizable alluvial plain of the Seyhan River. Adana, with a population of 1,131,000, is the region's economic heart, with industry based primarily on the production and processing of agricultural products, especially cotton. Agriculture has always played a rather important role in the development of this area's economy. Today, however, the economy is becoming more diversified. A large tanker terminal at Ceyhan provides a point of passage for trade, and the NATO air force base at Incirlik, where many U.S. military personnel are stationed, also provides an economic base. Antalya lies near the country's border with Syria. With a population of just over 600,000, it is the region's only large urban center. It is Turkey's leading port on the Mediterranean and also is a modern holiday resort and the southeastern gateway to the rest of the region.

CENTRAL ANATOLIA

The relocation of the Turkish capital city from the affluent Marmara Region to the interior was one of Atatürk's wisest

Anchored *gulets* (boats) can be seen through the window of an ancient fortress at Kal on the Turquoise Coast.

decisions. The move represented his bold attempt to connect the European and Asian portions of Turkey. The Central Anatolia Region (and the rest of Anatolia, for that matter) has long been an economic, population, and cultural backwater. It was a region of traditional people and cultural practices, relying primarily on farming and nomadic herding for economics. Poor transportation and communication linkages contributed to the region's isolation from the rest of the country. Surrounded by more prosper regions, Central Anatolia was, and still is, an area of little economic growth and heavy emigration of its people.

After Ankara (population 3,204,000) became the country's administrative center, considerable modernization has taken place. From its small-town roots, Ankara rose to become the second largest city in Turkey and a place of rapid urbanization and expansion. For that reason, and contrary to the rest of the country, few features of its landscape are of historical value. Its wide boulevards, recently built government and other buildings, and numerous museums convey the impression of a modern, cosmopolitan city. To show gratitude toward the father of the Turkish Republic, Mustafa Kemal (Atatürk), the Turks have built Anit Kabir, a remarkable mausoleum in the heart of the city. In addition to its role as the country's administrative center, Ankara plays a significant role in the heartland's rapidly diversifying and improving economy.

Other major towns are Konya (743,000) and Sivas (252,000). These communities owe their current status more to historical circumstances (they were important Seljuk settlements) than to other factors. The rest of Central Anatolia supports a much lower population density, with mostly small, rural settlements. Konya, the old Seljuk capital, is situated on ancient caravan routes. It is best known for being the center of the Sufi religious sect and associated events and activities, including "Whirling Dervishes." An interesting cultural event takes place each year in Konya. Each year, Dervishes—members of the religious order Mevlana, established in the thirteenth century by Celaleddin Rumi, who was known as Mevlana—celebrate his death (in 1273) by whirling around in the ecstasy of dance and music. This event was outlawed in the 1920s, as Turkey attempted to modernize. Now, however, the Dervishes are allowed to perform once a year.

EASTERN ANATOLIA

Toward the east and Turkey's borders with Armenia, Azerbaijan, and Iran, the landscape changes abruptly. Here,

Whirling dervishes are associated with the mystical Sufi religious sect.

rolling hills and plateaus and a steppe ecosystem give way to towering mountain peaks interrupted with deep canyons and river valleys. This is the Eastern Anatolian Region. Because of the rugged terrain and limited transportation routes, it is a region of low population density and considerable cultural isolation. Larger communities are found primarily in the valleys of the upper Tigris, Aras, and Murat rivers and in the vicinity of the Lake Van. Although Armenians consider this part of Turkey to be their homeland, it is now populated mostly by Kurds and Turks.

The main economic activity is agriculture. Transportation and communication networks are weak. Few roads and railroads connect provincial centers with the rest of the country, and most roads that do exist as links to the outside world are unpaved. On the other hand, because of its spectacular terrain, quaint villages, and traditional lifestyles, Eastern Anatolia has the potential of future development as a tourist destination. In absence of major economic development, there are few large cities. The region's leading urban areas are Malatya (381,000), Erzurum (361,000), Van (284,000), and Elazig (266,000).

SOUTHEASTERN ANATOLIA

Southeastern Anatolia is the final chapter in our overview of Turkey's regions. Many geographers simply include this region as a part of Eastern Anatolia. In this book, it is considered a separate region for cultural reasons. The area represents an extension of Arabic cultural intrusion into Turkey. Culturally, the region is part of Kurdistan—the territory settled by Kurds that includes portions of several other Southwest Asian countries. Politically, however, the area settled by Kurds in Southeastern Anatolia remains firmly in Turkish hands. This status is unlikely to change soon.

Diyarbakir (546,000), the unofficial Kurdish capital, is the region's most important (although not its largest) urban center. The city is situated on a major tributary of the Tigris River, astride the ancient trading route connecting the Arabian Peninsula and Asia Minor. In addition to being the regional Kurdish cultural center, Diyarbakir proudly offers a rich history. One highly visible feature of its colorful past is a three mile long city wall built by the Romans. Agriculture dominates as the economic activity of Southeastern Anatolia. During recent decades, however, the regional economy has been given a boost by increased textile manufacturing, construction, industrial diversification, and irrigation projects. Other cities in the region include Gaziantep (854,000) and Sanliurfa (386,000).

8

Turkey's Future

The author hopes that, after reading this book, you find Turkey to be a fascinating country worthy of further exploration. Turkey is not just another Asian country that warrants little more attention than that provoked by stories broadcast on the six o'clock news. Unfortunately, many people not knowledgeable of geography hold misleading stereotypes and false images of this wonderful country. For example, being close to Iraq and other Middle Eastern battlegrounds often casts Turkey in a negative light to potential travelers.

As is true of other developing (and developed) countries, Turkey has many problems. It also has much to offer. Perhaps the chief message of this book is that every culture has much to offer and is worthy of our exploration and interest. The first stage of such a journey always starts with reading material on a specific culture and otherwise becoming familiar with a way of life different from our own.

Many travelers to Turkey who want to experience and enjoy the richness of Turkic culture make their way to the carpet sellers and other purveyors of fine goods in the Grand Bazaar in Istanbul.

What are Turkey's prospects for the rest of the twenty-first century? In times of a downturn in the economy, it is difficult to be optimistic. In the long term, however, the country has many reasons for optimism. During the 1990s, the national economy expanded and living standards improved. Toward the end of the decade, however, the global economy slowed and Turkey suffered like many of the world's countries, including the United States. Almost certainly, once the global economy recovers, Turkey will again be on the road to prosperity. Becoming a member of the European Union (EU), a long-time Turkish dream, will certainly shorten the trip to economic growth. The EU has only a few reasons for not honoring Turkey with full membership, and the rest of Eastern Europe is gaining acceptance. After all, the rapidly aging European labor force will benefit from millions of work-ready Turks, just as it did from German workers in the past. The fear of some Western politicians that Turkey may fail to preserve itself as an officially secular state have so far proved to be unjustified. This concern appears to be the chief stumbling block in the way of acceptance into the EU membership.

On the domestic front, modernization remains a work in progress and one with a ways to go. Huge investments must be made in further developing the country's transportation and communication networks. Walking the crowded streets of Istanbul, Ankara, or Izmir is a wonderful experience for a traveler wanting to experience and enjoy the elements of Turkish culture. Unfortunately, one negative feature of that culture is poor urban planning and construction. Turkey is prone to severe earthquakes and the frequent resulting loss of property and life. To prevent such losses, urban planning must become a priority and building construction must be better able to withstand seismic shocks.

In terms of education, particularly the education of girls and young women, Turkey has achieved respectable success. This is especially true when compared to other dominantly

Muslim countries. As was noted, officially Turkey is a secular state, and the country's administration follows Atatürk's land-mark reforms. Despite administrative decisions, real change always comes slowly in a traditional society. For example, it has been difficult to change the ancient habit of women having limited opportunities in Turkey's male-dominated society. Gender equity must become a reality in all aspects of social, business, and political life. Turkey has already made positive steps in this direction. You will recall that the country has already had a woman as a leader of national government.

Turkey also must solve its ethnic problems before they result in further deterioration of the country's international image. Unless ethnic harmony is accomplished soon, all future political and economic improvements will be jeopardized. Kurds represent 20 percent of the country's population, and neither their numbers nor their plight can be ignored. Today, many of their demands sound unrealistic. Good solutions always include a two-way relationship in which parties are willing to compromise, even when smoldering issues from the past still linger.

Finally, as the country and its people lean increasingly toward the influence of the Western World, Turkey must not lose sight of its deeply entrenched geographical, historical, and cultural roots. A huge portion of the country, after all, lies on the Asian side of the Bosporus, Dardanelles, and Sea of Marmara. There is no reason why Turkey should be ashamed of belonging to the Middle East. The country has the maturity and leadership to play a major role in working toward future regional economic and political stability. Despite many obstacles, Turkey has proved that with innovative ideas, stable leadership, and controlled modernization, a nation can accomplish much. Certainly this land astride two continents can use much from its past in charting the route to a stable and prosperous future.

Fact at a Glance

Country name	Conventional long form: Republic of Turkey Conventional short form: Turkey Local short form: Turkiye
Government type	Republican parliamentary democracy
Independence	October 29, 1923 (Successor state to the Ottoman Empire)
Capital	Ankara
Area	Total: 301,389 square miles (780,594 square kilometers) Water: 3,791 square miles (9,819 square kilometers) Land: 297,597 square miles (770,773 square kilometers) Slightly larger than Texas
Land boundaries	Total: 1,645 miles (2,647 kilometers)
Border countries	Border countries: Syria, 510 miles (821 kilometers); Iraq, 218 miles (351 kilometers); Iran, 310 miles (500 kilometers); Azerbaijan, 5.59 miles (9 kilometers); Armenia, 166 miles (267 kilometers); Georgia, 156 miles (251 kilometers); Bulgaria, 149 miles (240 kilometers); Greece, 128 miles (206 kilometers)
Administrative divisions	81 provinces
Climate	Temperate; hot, dry summers with mild, wet winters; harsher in interior
Highest point	Mount Ararat, 16,948 feet (5,166 meters)
Coastline	4,474 miles (7,200 kilometers)
Arable land	Arable land: 34.53%, Permanent crops: 3.36% Other: 62.11% (1998 est.)
Population	67,308,928 (July 2002 est.) Males: 34,018,806 (51%) Females: 33,290,122 (49%)
Life expectancy at birth	Total population: 72 years Male: 69 years Female: 74 years
Literacy	Definition: age 15 and older can read and write Total population: 85% Males: 94% Females: 77%

Ethnic groups	Turkish 80%, Kurdish 20%
Religions	Muslim 99.8% (mostly Sunni), other 0.2% (mostly Christians and Jews)
GDP purchasing power parity (PPP)	$468 billion (2002 est.)
GDP per capita (PPP)	$7,000 (2002 est.)
Exports	$35.3 billion (2001)
Imports	$39.7 billion (2001)
Leading trade partners	Exports: Germany 17.2%, United States 10.0%, Italy 7.5%, United Kingdom 6.9%, France 6.0%, Russia 2.9% (2001)
	Imports: Germany 12.9%, Italy 8.4%, Russia 8.3%, United States 7.9%, France 5.5%, United States 4.6% (2001 est.)
Industries	Textiles, food processing, autos, mining (coal, chromite, copper, boron), steel, petroleum, construction, lumber, paper
Transportation	Railroads: 5,344 miles (8,633 kilometers), 1,305 miles (2,100 kilometers) electrified
	Highways: 237,374 miles (382,016 kilometers), 66,374 miles (106,819 kilometers) paved Waterways: 745 miles (1,199 kilometers) Airports: 120, 86 with paved runaways
Communications	TV Stations: 635 (1995) Phones (including cellular): about 40 million Internet Users: 2.5 million (2002)

B.C.

6,000+   Permanent settlements are established in Asia Minor.

Third millennium   Civilization was certainly present on the Aegean Sea and the Sea of Marmara shores, because radio carbon dating of the ancient city of Troy confirmed its origins at least to that period. Hurrians are settling in Anatolia.

Eighteenth – twelfth centuries   Hittites build a kingdom in Central Anatolia that will ultimately become one of the greatest powers of the ancient world.

Thirteenth century   Battle of Qadesh between Egyptians and Hittites.

Eleventh – seventh centuries   Phrygians control western Asia Minor.

Eighth century   Greek colonies spread on the shores of Asia Minor.

Seventh – sixth centuries   Lydia replaces Phrygia as the main power in the region.

Sixth century   Colonists from the Greek city Megara establish the town Byzantium on the European side of Bosphorus.

Sixth – fourth centuries   Persia controls Asia Minor until Alexander the Great's conquests in the 330s.

First century   The Roman Empire domination spreads across present-day Turkey.

A.D.

Fourth century   Emperor Constantine relocates the capital into the little known town of Byzantium, which is renamed Constantinople (Constantinopolis). The Roman Empire is divided into the Eastern and Western Roman Empires. All Asian Provinces now belong to the Eastern Empire.

476   The end of the Western Roman Empire. The Eastern Roman Empire will continue to exist until 1453.

527 – 565   Justinian I reigns. During this time, Constantinople sees astonishing architectural solutions, including the famous Aya Sophia church. Justinian I brought back the Empire's past glory.

Seventh–twelfth centuries	The Empire comes under attack from Arabs and Seljuks in interchanges of strong and weak rulers.
Ninth century	Seljuk Turks move to Mesopotamia to work as mercenaries for Bagdhad's caliph.
1055	Seljuks occupy Baghdad and soon control Christ's grave in Jerusalem as well.
1071	The Eastern Roman Empire loses to the Turks in the battle at Manzikert. This event marks the permanent Turkish presence in Anatolia.
1099	Crusaders enter Jerusalem in the First Crusades.
1204	Crusaders invade Constantinople and establish the Latin kingdom.
1204–1261	The Latin Kingdom is eliminated and Emperors regain control over Constantinople.
Thirteenth century	Osman Turks emerge as a cohesive nomadic group in Anatolia. They are named after their leader Osman (1284–1324).
Fourteenth– fifteenth centuries	Turks become the strongest power in Asia Minor, reducing the size of the Roman Empire to Constantinople and surroundings while establishing an empire themselves.
1396	The Battle at Nicopolis occurs.
1402	Timur's forces invade Anatolia.
1453	Turks capture Constantinople.
1520–1566	Süleyman the Magnificent reigns.
1683–1699	The Ottoman Empire is at war with the Habsburg Monarchy and the empire's slow decline begins.
Eighteenth– nineteenth centuries	The Ottoman Empire's power further declines, and provinces ask for independence.
1878	Congress in Berlin marks the loss of almost all European possessions.
1914–1918	The Ottoman Empire enters World War I on the German side and loses most of its prewar provinces.

1919	The Treaty of Versailles ends the war.
1919–1922	Turkey resists foreign occupation and the Ottoman Empire continues under the leadership of Mustafa Kemal (Atatürk).
1923	The Republic of Turkey becomes independent.
1938	Atatürk dies.
1945	Turkey enters World War II on the side of the Allies and becomes a founding member of The United Nations.
1952	Turkey enters NATO.
1960, 1971, 1980	Major interventions of military forces into political life of Turkey bring governmental changes.
1974	Turkey invades of northern Cyprus.
1984	The armed Kurdish rebellion begins.
1993	Tansu Ciller becomes the first woman prime minister.
1999	Kurdish leader Abdullah Ocalan is captured.
2003	Tayyip Erdogan becomes a prime minister and forms a government.

Ceram, C.W. *The Secret of the Hittites: The Discovery of an Ancient Empire.* New York: Alfred A. Knopf, 1956.

Davison, Roderic H. *Turkey.* Englewood Cliffs, NJ: Prentice-Hall, Inc., 1968.

Dewdney, Joe C. *Turkey: An Introductory Geography.* New York: Praeger Publishers, 1971.

Downey, Glanville. *Constantinople in the Age of Justinian.* Norman: University of Oklahoma Press, 1960.

Feinstein, Stephen C. *Turkey: In Pictures.* Minneapolis, MN: Lerner Publishing Group, 2003.

Goodwin, Jason. *Lords of the Horizons: A History of the Ottoman Empire.* New York: Picador USA, 2002.

Lewis, Bernard. *The Emergence of Modern Turkey.* New York: Oxford University Press, 2002.

Macqueen, J. G. *The Hittites and Their Contemporaries in Asia Minor.* Boulder, CO: Westview Press, 1975.

Mayne, Peter. *Cities of the World: Istanbul.* New York: A. S. Barnes, 1967.

Pierce, J. E. *Life in a Turkish Village.* New York: Holt, Rinehart and Winston, 1964.

Stewart, Desmond. *Turkey.* New York: Time-Life Books, 1969.

Stoneman, Richard. *A Traveler's History of Turkey.* New York: Interlink Books, 1993.

Yenen, Serif. *Turkish Odyssey: A Cultural Guide to Turkey.* London: Milet Publishing, 2003.

Index

Index

Index

page:

11: © Craig Lovell/CORBIS
13: © Lucidity Information Design, LLC
17: © Adam Woolfitt/CORBIS
19: © Lucidity Information Design, LLC
23: © Yann Artus-Bertrand/CORBIS
25: © Adam Woolfitt/CORBIS
28: © Craig Lovell/CORBIS
52: Associated Press, AP
57: Courtesy CIA

60: Associated Press, AP/Musrad Sezer
67: Associated Press, AP/Burhan Ozbilici
71: Associated Press, AP/Murad Sezer
77: Associated Press, AP/Burhan Ozbilici
79: Associated Press, AP/David Karp
84: © Arthur Thévenart/CORBIS
93: © Patrick Ward/CORBIS
98: © Craig Lovell/CORBIS
103: Associated Press, AP/Murad Sezer

Cover: © O. Alamany & E. Vicens/CORBIS

ZORAN "ZOK" PAVLOVIĆ is a professional geographer who resides and works in Brookings, South Dakota. His previous contributions to the Chelsea House's "Major World Nations" were *The Republic Of Georgia* (with Charles F. "Fritz" Gritzner), *Croatia*, and *Kazakhstan*. When not studying and writing, Zok enjoys motorcycle traveling and gourmet cooking. He visited Italy numerous times, the latest visit being an adventure through the central regions in spring of 2003.

CHARLES F. ("FRITZ") GRITZNER is Distinguished Professor of Geography at South Dakota University in Brookings. He is now in his fifth decade of college teaching and research. During his career, he has taught more than 60 different courses, spanning the fields of physical, cultural, and regional geography. In addition to his teaching, he enjoys writing, working with teachers, and sharing his love for geography with students. As consulting editor for the MODERN WORLD NATIONS series, he has a wonderful opportunity to combine each of these "hobbies." Fritz has served as both President and Executive Director of the National Council for Geographic Education and has received the Council's highest honor, the George J. Miller Award for Distinguished Service.